Scikit-Learn 机器学习
核心技术与实践

谭贞军 ◎编著

清华大学出版社
北京

内 容 简 介

本书循序渐进地讲解了使用 Scikit-Learn 开发机器学习程序的核心知识，并通过具体实例的实现过程演练了使用 Scikit-Learn 的方法和流程。全书共 10 章，包括人工智能与 Scikit-Learn 简介，加载数据集，监督学习，无监督学习，模型选择和评估，数据集转换，实现大数据计算，英超联赛比分预测系统（Matplotlib+Scikit-Learn+Flask+Pandas），AI 考勤管理系统（face-recognition+Matplotlib+Django+Scikit-Learn+Dlib），实时电影推荐系统（Scikit-Learn+Flask+Pandas）。本书简洁而不失其技术深度，内容丰富全面，易于阅读。

本书适用于已经了解 Python 语言基础语法的读者，以及想进一步学习机器学习和深度学习技术的读者，还可以作为大专院校相关专业的师生用书和培训学校的专业教材。

本书封面贴有清华大学出版社防伪标签，无标签者不得销售。

版权所有，侵权必究。举报：010-62782989，beiqinquan@tup.tsinghua.edu.cn。

图书在版编目(CIP)数据

Scikit-Learn 机器学习核心技术与实践 / 谭贞军编著. —北京：清华大学出版社，2022.3

ISBN 978-7-302-59948-7

Ⅰ. ①S… Ⅱ. ①谭… Ⅲ. ①机器学习 Ⅳ. ①TP181

中国版本图书馆 CIP 数据核字 (2022) 第 022298 号

责任编辑：魏　莹
封面设计：李　坤
责任校对：李玉茹
责任印制：沈　露

出版发行：清华大学出版社
　　　网　　址：http://www.tup.com.cn，http://www.wqbook.com
　　　地　　址：北京清华大学学研大厦 A 座　　邮　编：100084
　　　社 总 机：010-83470000　　　　　　　　　邮　购：010-62786544
　　　投稿与读者服务：010-62776969，c-service@tup.tsinghua.edu.cn
　　　质 量 反 馈：010-62772015，zhiliang@tup.tsinghua.edu.cn
印 装 者：北京嘉实印刷有限公司
经　　销：全国新华书店
开　　本：185mm×230mm　　　印　张：17.5　　　字　数：431 千字
版　　次：2022 年 3 月第 1 版　　印　次：2022 年 3 月第 1 次印刷
定　　价：79.00 元

————————————————————————————————

产品编号：091674-01

前 言

人工智能就是我们平常所说的 AI，全称是 Artificial Intelligence。人工智能是研究、开发用于模拟、延伸和扩展人类智能的理论、方法、技术及应用系统的一门新的技术科学。Scikit-Learn 也被称为 sklearn，是针对 Python 编程语言的免费软件机器学习库。Scikit-Learn 具有各种分类、回归和聚类算法，包括支持向量机、随机森林、梯度提升、K-均值和 DBSCAN，旨在与 Python 数值科学库 NumPy 和 SciPy 联合使用。

（一）本书特色

1. 内容全面

本书详细讲解了使用 Scikit-Learn 开发人工智能程序的技术知识，循序渐进地讲解了这些技术的使用方法和技巧，帮助读者快速步入 Python 人工智能开发高手之列。

2. 实例驱动教学

本书采用理论加实例的教学方式，通过对这些实例的分析，实现了对知识点的横向切入和纵向比较，让读者有更多的实践演练机会，并且可以从不同的方位展现一个知识点的用法，真正实现了拔高的教学效果。

3. 详细介绍了 Scikit-Learn 开发的流程

本书从一开始便对 Scikit-Learn 开发流程进行了详细介绍，而且在讲解中结合了多个实用性很强的数据分析项目案例，带领读者掌握 Scikit-Learn 开发的相关知识，以解决实际工作中的问题。

4. 扫描二维码，获取配书学习资源

本书正文的每个二级标题后都放了一个二维码，读者可通过扫描二维码在线观看讲解视频，其中既包括实例讲解，也包括教程讲解。此外，读者还可以扫描右侧的二维码获取书中案例源代码。

扫码下载全书源代码

（二）本书内容

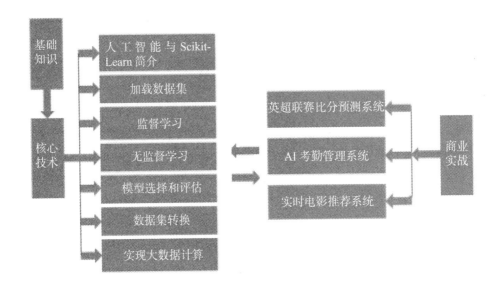

（三）本书读者对象

（1）软件研发工程师；
（2）Python 语言初学者；
（3）机器学习开发人员；
（4）数据库工程师和管理员；
（5）中学及大学教育工作者。

（四）致谢

本书在编写过程中，首先，感谢清华大学出版社编辑老师的大力支持，正是他们的求实、耐心和高效工作，才使本书能够顺利出版。其次，十分感谢我的家人给予的巨大支持。因作者水平有限，书中存在纰漏之处在所难免，恳请读者提出宝贵的意见或建议，以便修订使之更臻完善。

编　者

目 录

第 1 章 人工智能与Scikit-Learn 简介 ··· 1

1.1 人工智能技术的兴起 ····················· 2
- 1.1.1 人工智能介绍 ····················· 2
- 1.1.2 人工智能的研究领域 ········· 2
- 1.1.3 和人工智能相关的几个重要概念 ································· 3
- 1.1.4 人工智能的两个重要发展阶段 ································· 4

1.2 机器学习和深度学习 ····················· 4
- 1.2.1 机器学习 ····························· 5
- 1.2.2 深度学习 ····························· 5
- 1.2.3 机器学习和深度学习的区别 ··· 6

1.3 初步认识 Scikit-Learn ················· 7
- 1.3.1 Scikit-Learn 介绍 ············· 7
- 1.3.2 使用 pip 安装 Scikit-Learn ··· 7
- 1.3.3 使用 Anaconda 安装 Scikit-Learn ································· 8
- 1.3.4 解决速度过慢的问题 ········· 9

1.4 准备开发工具 ································· 10

第 2 章 加载数据集 ···················· 11

2.1 标准数据集 API ····························· 12
- 2.1.1 波士顿房价数据集（适用于回归任务）······························· 12
- 2.1.2 威斯康星州乳腺癌数据集（适用于分类问题）················ 13
- 2.1.3 糖尿病数据集（适用于回归任务）····························· 13
- 2.1.4 手写数字数据集（适用于分类任务）····························· 14
- 2.1.5 Fisher 的鸢尾花数据集（适用于分类问题）················ 15
- 2.1.6 红酒数据集（适用于分类问题）··································· 15

2.2 自定义数据集 ································· 16
- 2.2.1 生成聚类数据 ··················· 16
- 2.2.2 生成同心圆样本点 ··········· 17
- 2.2.3 生成模拟分类数据集 ······· 18
- 2.2.4 生成太极型非凸集样本点 ··· 19

第 3 章 监督学习 ······················· 21

3.1 广义线性模型 ································· 22
- 3.1.1 普通最小二乘法 ··············· 22
- 3.1.2 岭回归 ······························· 24
- 3.1.3 Lasso 回归 ························· 25

3.2 线性判别分析和二次判别分析 ··· 27
- 3.2.1 使用线性判别分析来降维 ··· 28
- 3.2.2 LDA 和 QDA 分类器的数学公式 ································· 30
- 3.2.3 收缩 ··································· 30

3.3 内核岭回归 ………………… 32

3.4 支持向量机 ………………… 33
 3.4.1 分类 ………………… 34
 3.4.2 回归 ………………… 36
 3.4.3 密度估计和异常检测 ……… 38

3.5 随机梯度下降 ……………… 39
 3.5.1 分类 ………………… 40
 3.5.2 回归 ………………… 42
 3.5.3 稀疏数据的随机梯度下降 … 42

第 4 章 无监督学习 …………… 49

4.1 高斯混合模型 ……………… 50
 4.1.1 高斯混合 ……………… 50
 4.1.2 变分贝叶斯高斯混合 …… 53

4.2 流形学习 …………………… 55

4.3 聚类 ………………………… 57
 4.3.1 KMeans 算法 ………… 57
 4.3.2 MiniBatchKMeans 算法 … 60

4.4 双聚类 ……………………… 62
 4.4.1 谱聚类算法 …………… 63
 4.4.2 光谱联合聚类算法 ……… 66

第 5 章 模型选择和评估 ………… 69

5.1 交叉验证：评估估算器的表现 … 70
 5.1.1 计算交叉验证的指标 …… 71
 5.1.2 交叉验证迭代器 ………… 75

5.2 调整估计器的超参数 ………… 76
 5.2.1 网格追踪法：穷尽的网格搜索 ………………… 77
 5.2.2 随机参数优化 ………… 80

5.3 模型评估：量化预测的质量 …… 82
 5.3.1 得分参数 scoring：定义模型评估规则 ………… 83
 5.3.2 分类指标 ……………… 85

第 6 章 数据集转换 ……………… 89

6.1 Pipeline（管道）和 FeatureUnion（特征联合）………………… 90
 6.1.1 Pipeline：链式评估器 …… 90
 6.1.2 FeatureUnion（特征联合）：特征层 ………………… 93

6.2 特征提取 …………………… 95
 6.2.1 从字典类型加载特征 …… 95
 6.2.2 特征哈希 ……………… 96
 6.2.3 提取文本特征 ………… 97
 6.2.4 提取图像特征 ………… 99

6.3 预处理数据 ·················· 102
    6.3.1 标准化处理 ············· 102
    6.3.2 非线性转换 ············· 103

6.4 无监督降维 ·················· 106
    6.4.1 PCA：主成分分析 ······· 106
    6.4.2 随机投影 ··············· 110

第 7 章 实现大数据计算 ········· 117

7.1 计算扩展策略 ················ 118
    7.1.1 使用外核学习实例进行
          拓展 ···················· 118
    7.1.2 使用外核方法进行分类 ···· 119

7.2 计算性能 ···················· 129
    7.2.1 预测延迟 ··············· 129
    7.2.2 预测吞吐量 ············· 137

第 8 章 英超联赛比分预测系统（Matplotlib+Scikit-Learn+Flask+Pandas）········ 143

8.1 英超联赛介绍 ················ 144

8.2 系统模块介绍 ················ 144

8.3 数据集 ······················ 144
    8.3.1 获取 api-football 密钥 ······ 145
    8.3.2 获取数据 ··············· 145
    8.3.3 收集最新数据 ··········· 150

8.4 特征提取和数据可视化 ········ 152
    8.4.1 提取数据 ··············· 153
    8.4.2 数据可视化 ············· 154

8.5 模型选择和训练 ·············· 161
    8.5.1 机器学习函数 ··········· 161
    8.5.2 数据降维 ··············· 172
    8.5.3 MLP 神经网络 ··········· 175

8.6 模型评估 ···················· 178
    8.6.1 近邻模型和混淆矩阵模型 ··· 179
    8.6.2 随机森林模型和混淆矩阵
          模型 ···················· 186
    8.6.3 SVM 模型和混淆矩阵
          模型 ···················· 190

8.7 Web 可视化 ·················· 197
    8.7.1 获取预测数据 ··········· 197
    8.7.2 Flask Web 主页 ·········· 200

第 9 章 AI 考勤管理系统（face-recognition+Matplotlib+Django+Scikit-Learn+Dlib）·············· 203

9.1 背景介绍 ···················· 204

9.2 系统需求分析 ················ 204
    9.2.1 可行性分析 ············· 204

9.2.2 系统操作流程分析 ············ 204
9.2.3 系统模块设计 ············ 204

9.3 系统配置 ························ 205
9.3.1 Django 配置文件 ············ 205
9.3.2 路径导航文件 ············ 206

9.4 用户注册和登录验证 ············ 207
9.4.1 登录验证 ············ 207
9.4.2 添加新用户 ············ 208
9.4.3 设计数据模型 ············ 210

9.5 采集照片和机器学习 ············ 210
9.5.1 设置采集对象 ············ 210
9.5.2 采集照片 ············ 212
9.5.3 训练照片模型 ············ 214

9.6 考勤打卡 ························ 216
9.6.1 上班打卡签到 ············ 216
9.6.2 下班打卡 ············ 218

9.7 可视化考勤数据 ················ 220
9.7.1 统计最近两周员工的考勤数据 ············ 220
9.7.2 查看本人在指定时间段内的考勤信息 ············ 225

9.7.3 查看某员工在指定时间段内的考勤信息 ············ 229

第 10 章 实时电影推荐系统（Scikit-Learn+Flask+Pandas）··· 233

10.1 系统介绍 ························ 234
10.1.1 背景介绍 ············ 234
10.1.2 推荐系统和搜索引擎 ············ 234
10.1.3 项目介绍 ············ 235

10.2 系统模块 ························ 235

10.3 数据采集和整理 ················ 235
10.3.1 数据整理 ············ 236
10.3.2 电影详情数据 ············ 239
10.3.3 提取电影特征 ············ 247

10.4 情感分析和序列化操作 ········ 252

10.5 Web 端实时推荐 ················ 253
10.5.1 Flask 启动页面 ············ 253
10.5.2 模板文件 ············ 256
10.5.3 后端处理 ············ 263

第 1 章

人工智能与 Scikit-Learn 简介

近年来，随着人工智能技术的飞速发展，机器学习和深度学习技术已经摆在了人们的面前，并成为程序员的学习热点。在本章的内容中，将详细介绍人工智能和 Scikit-Learn 基础知识，为读者步入本书后面知识的学习打下基础。

1.1 人工智能技术的兴起

在本节的内容中,将简要介绍人工智能技术的基本知识。

扫码观看本节视频讲解

1.1.1 人工智能介绍

人类文明发展几千年,自从机器诞生以来,聪明的人类就开始试图让机器具有智能,也就是人工智能。人工智能是一门极富挑战性的科学,从事这项工作的人必须懂得计算机知识、心理学和哲学。人工智能包含的内容十分广泛,它由不同的领域组成,如机器学习、计算机视觉等。总的来说,人工智能研究的一个主要目标是使机器能够胜任一些通常需要人类智能才能完成的复杂工作。

人工智能,单从字面上应该理解为人类创造的智能。那么什么是智能呢?如果人类创造了一个机器人,这个机器人能具有像人类一样甚至超过人类的推理、知识、学习、感知处理等这些能力,那么就可以将这个机器人称为是一个有智能的物体,也就是人工智能。

现在通常将人工智能分为弱人工智能和强人工智能。我们看到电影里的一些人工智能大部分都是强人工智能,他们能像人类一样思考如何处理问题,甚至能在一定程度上做出比人类更好的决定,他们能自适应周围的环境,解决一些程序中没有遇到的突发事件,具备这些能力的就是强人工智能。但是在目前的现实世界中,大部分人工智能只是实现了弱人工智能,能够让机器具备观察和感知能力,在经过一定的训练后能计算一些人类不能计算的东西,但是它并没有自适应能力,也就是说,它不会处理突发的情况,只能处理程序中已经写好的、已经预测到的事情,这就叫作弱人工智能。

1.1.2 人工智能的研究领域

人工智能的研究领域主要有五层,从下往上的具体说明如图1-1所示。
- 第1层:基础设施层,包含大数据和计算能力(硬件配置)两部分,数据越大,人工智能的能力越强。
- 第2层:算法层,例如卷积神经网络、LSTM序列学习、Q-Learning和深度学习等算法都是机器学习的算法。
- 第3层:技术方向层,例如计算机视觉、语音工程和自然语言处理等。另外,还有规划决策系统,例如增强学习(Reinforcement Learning),或类似于大数据分析的统计系统,这些都能在机器学习算法上产生。
- 第4层:具体技术层,例如图像识别、图像理解、语音识别、语义理解、情感分析、机器翻译等。
- 第5层:行业解决方案层,例如人工智能在金融、医疗、安防、交通、互联网和游戏等领域的应用。

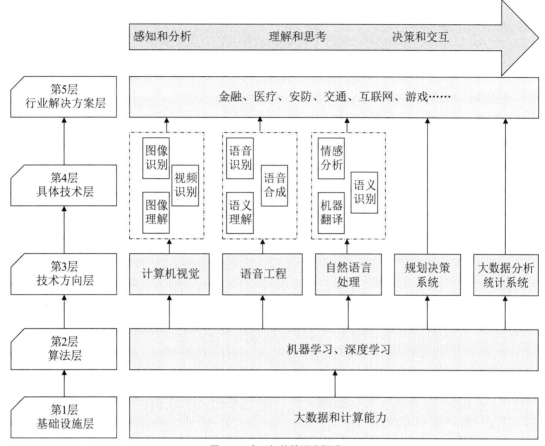

图 1-1 人工智能的研究领域

1.1.3 和人工智能相关的几个重要概念

1. 监督学习

监督学习的任务是学习一个模型，这个模型可以处理任意的一个输入，并且针对每个输入都可以映射输出一个预测结果。这个模型相当于数学中的一个函数，输入就相当于数学中的 X，而预测的结果就相当于数学中的 Y。对于每一个 X，我们都可以通过一个映射函数映射出一个结果。

2. 非监督学习

非监督学习是指直接对没有标记的训练数据进行建模学习，注意，在这里的数据是没有标记的数据，与监督学习的最基本的区别是建模的数据，一个有标签，一个没有标签。例如聚类（将物理或抽象对象的集合分成由类似的对象组成的多个类的过程）是一种典型的非监督学习，分类就是一种典型的监督学习。

3. 半监督学习

当我们拥有标记的数据很少，未被标记的数据很多，但是人工标注又比较昂贵的时候，可以根据一些条件（查询算法）查询（query）一些数据，让专家进行标记。这是半监督学习与其他算法的本质区别。所以说对主动学习的研究主要是设计一种框架模型，运用新的查询算法查询需要专家来认为标注的数据。最后用查询到的样本训练分类模型来提高模型的精确度。

4. 主动学习

当使用一些传统的监督学习方法做分类处理的时候，通常是训练样本的规模越大，分类的效果就越好。但是在现实生活的很多场景中，标记样本的获取是比较困难的，这需要领域内的专家来进行人工标注，所花费的时间成本和经济成本都是很大的。而且，如果训练样本的规模过于庞大，训练的时间花费也会比较多。那么问题来了：有没有一种有效办法，能够使用较少的训练样本来获得性能较好的分类器呢？主动学习（active learning）为我们提供了这种可能。主动学习通过一定的算法查询最有用的未标记样本，并交由专家进行标记，然后用查询到的样本训练分类模型来提高模型的精确度。

在人类的学习过程中，通常利用已有的经验来学习新的知识，又依靠获得的知识来总结和积累经验，经验与知识不断交互。同样，机器学习模拟人类学习的过程，利用已有的知识训练出模型去获取新的知识，并通过不断积累的信息去修正模型，以得到更加准确、有用的新模型。不同于被动学习被动地接受知识，主动学习能够选择性地获取知识。

1.1.4　人工智能的两个重要发展阶段

1）推理期

20 世纪 50 年代，人工智能的发展经历了"推理期"，通过赋予机器逻辑推理能力使机器获得智能，当时的 AI 程序能够证明一些著名的数学定理，但由于机器缺乏知识，远不能实现真正的智能。

2）知识期

20 世纪 70 年代，人工智能的发展进入"知识期"，即将人类的知识总结出来教给机器，使机器获得智能。在这一时期，大量的专家系统问世，在很多领域取得大量成果，但由于人类知识量巨大，故出现"知识工程瓶颈"。

1.2　机器学习和深度学习

在人工智能的两个发展阶段中，无论是"推理期"还是"知识期"，都会存在如下两个缺点。

（1）机器都是按照人类设定的规则和总结的知识运作，永远无法超越其创造者——人类。

扫码观看本节视频讲解

（2）人力成本太高，需要专业人才进行具体实现。

基于上述两个缺点，人工智能技术的发展出现了一个瓶颈期。为了突破这个瓶颈期，一些权威学者就想到，如果机器能够自我学习，问题不就迎刃而解了吗？此时机器学习（machine learning，ML）技术便应运而生，人工智能便进入"机器学习"时代。本节将简要介绍机器学习的基本知识。

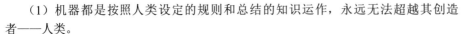

1.2.1 机器学习

机器学习是一门多领域交叉学科，涉及概率论、统计学、逼近论、凸分析、算法复杂度理论等多门学科。机器学习专门研究计算机怎样模拟或实现人类的学习行为，以获取新的知识或技能，重新组织已有的知识结构使之不断改善自身的性能。

机器学习是一类算法的总称，这些算法企图从大量历史数据中挖掘出其中隐含的规律，并用于预测或者分类。更具体地说，机器学习可以看作是寻找一个函数，输入是样本数据，输出是期望的结果，只是这个函数过于复杂，以至于不太方便形式化表达。需要注意的是，机器学习的目标是使学到的函数很好地适用于"新样本"，而不仅仅是在训练样本上表现很好。学到的函数适用于新样本的能力，称为泛化（generalization）能力。

机器学习有一个显著的特点，也是机器学习最基本的做法，就是使用一个算法从大量的数据中解析并得到有用的信息，接着从中学习，然后对之后真实世界中会发生的事情进行预测或作出判断。机器学习需要海量的数据来进行训练，并从这些数据中得到有用的信息，然后反馈到真实世界的用户中。

我们可以用一个简单的例子来说明机器学习，假设在天猫或京东购物的时候，天猫和京东会向消费者推送商品信息，这些推荐的商品往往是消费者很感兴趣的商品，这个过程是通过机器学习完成的。其实这些推送商品是天猫和京东根据消费者以前的购物订单和经常浏览的商品记录而得出的结论，可以从中得出商城中的哪些商品是消费者感兴趣的并且会大概率购买，然后将这些商品定向推送给消费者。

1.2.2 深度学习

前面介绍的机器学习是一种实现人工智能的方法，而深度学习是一种实现机器学习的技术。深度学习本来并不是一种独立的学习方法，其本身也会用到有监督和无监督的学习方法来训练深度神经网络。但由于近几年该领域发展迅猛，一些特有的学习手段相继被提出（如残差网络），因此越来越多的人将其单独看作一种学习的方法。

假设我们需要识别某个照片是狗还是猫，如果是传统机器学习的方法，会首先定义一些特征，如有没有胡须、耳朵、鼻子、嘴巴的模样等。总之，我们首先要确定相应的"面部特征"作为机器学习的特征，以此来对对象进行分类识别。深度学习的方法则更进一步，它会自动找出这个分类问题所需要的重要特征，而传统机器学习则需要人工给出特征。那么，深度学习是如何做到这一点的呢？下面继续以猫、狗识别的例子进行说明。

（1）确定有哪些边和角与识别出猫、狗的关系最大。
（2）根据上一步找出的很多小元素（边、角等）构建层级网络，找出它们之间的各种组合。
（3）在构建层级网络之后，就可以确定哪些组合可以识别出猫和狗。

⚠️ **注 意** 其实深度学习并不是一个独立的算法，在训练神经网络的时候通常也会用到监督学习和无监督学习。但是由于一些独特的学习方法被提出，把它看成单独的一种学习的算法也可以。深度学习可以大致理解成包含多个隐含层的神经网络结构，深度学习的"深"字指的就是隐藏层的深度。

1.2.3 机器学习和深度学习的区别

在机器学习方法中,几乎所有的特征都需要通过行业专家先确定,然后手工对特征进行编码,而深度学习算法会自己从数据中学习特征。这也是深度学习十分引人注目的一点,毕竟特征工程是一项十分烦琐、耗费很多人力物力的工作,深度学习的出现则大大减少了发现特征的成本。

在解决问题时,传统机器学习算法通常先把问题分成几块,一个个地解决好之后,再重新组合起来。但是深度学习则是一次性地、端到端地解决。假如存在一个任务:识别出在某图片中有哪些物体,并找出它们的位置。

传统机器学习的做法是把问题分为两步:发现物体和识别物体。首先,我们用几个物体边缘的盒型检测算法,把所有可能的物体都框出来。然后,再使用物体识别算法,识别出这些物体分别是什么。图1-2是一个机器学习识别的例子。

图 1-2　机器学习的识别

但是深度学习不同,它会直接在图片中把对应的物体识别出来,同时还能标明对应物体的名字。这样就可以做到实时识别物体,例如,YOLO net 可以在视频中实时识别物体,图1-3是YOLO在视频中实现深度学习识别的例子。

图 1-3　深度学习的识别

⚠ **注意**　人工智能、机器学习、深度学习三者的关系

机器学习是实现人工智能的方法;深度学习是机器学习的一种算法,一种实现机器学习的技术和学习方法。

1.3 初步认识 Scikit-Learn

Scikit-Learn 具有各种分类、回归和聚类算法,包括支持向量机、随机森林、梯度提升、k 均值和 DBSCAN,并且旨在与 Python 数值科学库 NumPy 和 SciPy 联合使用。

扫码观看本节视频讲解

1.3.1 Scikit-Learn 介绍

Scikit-Learn 主要是用 Python 语言编写的,并且广泛使用 NumPy 进行高性能的线性代数和数组运算。Scikit-Learn 与许多其他 Python 库很好地集成在一起,例如 Matplotlib 和 Plotly 用于绘图,NumPy 用于数组矢量化,Pandas 用于处理数据帧,SciPy 用于数学处理等。

Scikit-Learn 最初由 David Cournapeau 于 2007 年在 Google(谷歌)的夏季代码项目中开发。后来 Matthieu Brucher 加入该项目,并开始将其用作论文工作的一部分。2010 年,法国计算机科学与自动化研究所(INRIA)参与其中,并于 2010 年 1 月下旬发布了第一个公开版本(v0.1 beta)。

Scikit-Learn 是基于第三方库 NumPy、SciPy 和 Matplotlib 实现的,这三个库的具体说明如下。

- NumPy:Python实现的开源科学计算包,可以定义高维数组对象,提供了矩阵计算和随机数生成等内置函数。
- SciPy:Python实现的高级科学计算包。它和NumPy联系很密切,SciPy一般都是操控NumPy数组来进行科学计算,所以可以说是基于NumPy之上了。SciPy有很多子模块可以应对不同的应用,例如插值运算、优化算法、图像处理、数学统计等。
- Matplotlib:Python实现的作图包。使用Matplotlib能够非常简单地可视化数据,仅需要几行代码,便可以生成直方图、功率谱、条形图、错误图、散点图等。

1.3.2 使用 pip 安装 Scikit-Learn

安装 Scikit-Learn 最简单的方法是使用 pip 命令。在使用这种安装方式时,无须考虑当前所使用的 Python 版本和操作系统的版本,pip 会自动安装适合用户当前 Python 版本和操作系统版本的 Scikit-Learn。在安装 Python 后,会自动安装 pip。

(1)在 Windows 系统中单击左下角的▓图标,在弹出的菜单中找到"命令提示符"选项,然后用鼠标右键单击"命令提示符"选项,在弹出的快捷菜单中依次选择"更多"→"以管理员身份运行"命令,如图 1-4 所示。

图 1-4 以管理员身份运行命令提示符

（2）在弹出的命令提示符界面中输入如下命令即可安装库 Scikit-Learn：

```
pip install Scikit-learn
```

在输入上述 pip 安装命令后，会弹出下载并安装 Scikit-Learn 的界面，如图 1-5 所示。因为库 Scikit-Learn 的容量比较大，并且还需要安装相关的其他库，所以整个下载安装过程会比较慢，需要大家耐心等待，确保 Scikit-Learn 能够安装成功。

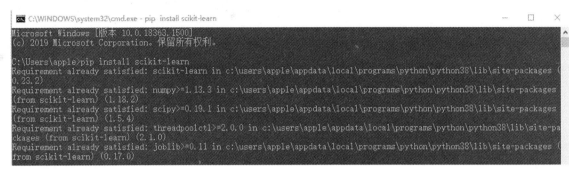

图 1-5 下载并安装 Scikit-Learn 界面

⚠️ **注 意** 使用 pip 命令安装的另一好处是，自动为用户安装适合的当前最新版本的 Scikit-Learn。因为在笔者电脑中安装的是 Python 3.8，并且操作系统是 64 位的 Windows 10，所以图 1-5 所示的是适合笔者的最新版本的安装文件：Scikit_learn-0.24.2-cp38-cp38-win_amd64.whl。在这个安装文件的名字中，各个字段的含义如下。

- scikit_learn-0.24.2：表示Scikit-Learn的版本号是0.24.2。
- cp38：表示适用于Python 3.8版本。
- win_amd64：表示适用于64位的Windows操作系统。

在使用前面介绍的 pip 方式下载安装 Scikit-Learn 时，能够安装成功的一个关键因素是网速。如果你的网速过慢，可以考虑在百度中搜索一个 Scikit-Learn 下载包。因为目前适合笔者的最新版本的安装文件是 scikit_learn-0.24.2-cp38-cp38-win_amd64.whl，那么笔者可以在百度中搜索这个文件，然后下载。下载完成后保存到本地硬盘中，例如，保存位置是：D:\scikit_learn-0.24.2-cp38-cp38-win_amd64.whl，在命令提示符界面中定位到 D 盘根目录，然后运行如下命令就可以安装 Scikit-Learn 了。

```
pip install scikit_learn-0.24.2-cp38-cp38-win_amd64.whl
```

1.3.3 使用 Anaconda 安装 Scikit-Learn

使用 Anaconda 安装 Scikit-Learn 的方法和上面介绍的 pip 方式相似，具体流程如下。

（1）在 Windows 系统中单击左下角的▓图标，然后选择 Anaconda Powershell Prompt 命令，在弹出的菜单中依次选择"更多"→"以管理员身份运行"命令，如图 1-6 所示。

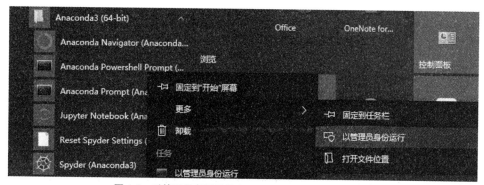

图 1-6　以管理员身份运行 Anaconda Powershell Prompt

（2）在弹出的命令提示符界面中输入如下命令即可安装库 Scikit-Learn：

```
pip install Scikit-Learn
```

在输入上述 pip 安装命令后，会弹出下载并安装 Scikit-Learn 的界面，安装成功后的界面效果如图 1-7 所示。

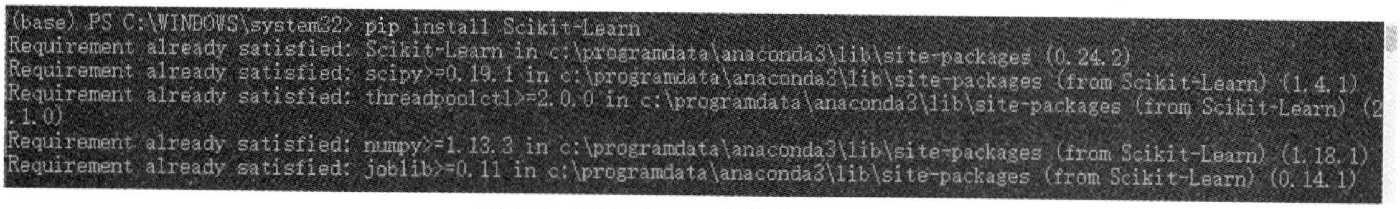

图 1-7　Scikit-Learn 安装成功后的界面

1.3.4　解决速度过慢的问题

在使用前面介绍的 pip 方式安装库 Scikit-Learn 时，经常会遇到因为网速过慢而安装失败的问题。这是因为库 Scikit-Learn 的安装包保存在国外的服务器中，所以国内用户在下载时会遇到网速过慢的问题。为了解决这个问题，国内很多网站也为开发者提供了常用的 Python 库的安装包，例如清华大学和豆瓣网等。

（1）使用清华源安装 Python 库的语法格式如下：

```
pip install -i https://pypi.tuna.tsinghua.edu.cn/simple 库的名字
```

例如，在 Windows 10 系统的命令提示符界面中，输入下面的命令即可安装 Scikit-Learn：

```
pip install -i https://pypi.tuna.tsinghua.edu.cn/simple Scikit-Learn
```

（2）使用豆瓣源安装 Python 库的语法格式如下：

```
pip install 库的名字 -i http://pypi.douban.com/simple/ --trusted-host pypi.douban.com
```

例如，在 Windows 10 系统的命令提示符界面中，输入下面的命令即可安装 Scikit-Learn：

```
pip install Scikit-Learn -i http://pypi.douban.com/simple/ --trusted-host pypi.douban.com
```

1.4 准备开发工具

对于 Python 开发者，建议使用 PyCharm 开发并调试运行 Scikit-Learn 程序。
（1）打开 PyCharm，然后新建一个名为 "first" 的 Python 工程，如图 1-8 所示。

扫码观看本节视频讲解

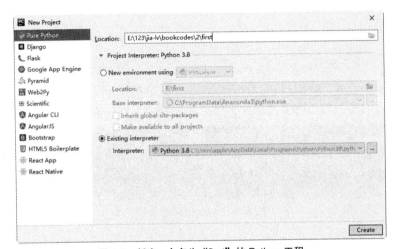

图 1-8 新建一个名为 "first" 的 Python 工程

（2）在工程 "first" 中新建一个 Python 程序文件 first.py，然后编写如下代码：

```
from sklearn import linear_model
reg = linear_model.LinearRegression()
print(reg.fit([[0, 0], [1, 1], [2, 2]], [0, 1, 2]))
print(reg.coef_)
```

上述代码的功能是，通过 LinearRegression 调用 fit() 方法来拟合数组 X、y，并且将线性模型的系数 w 存储在其成员变量 coef_ 中。在电脑中执行后会输出：

```
LinearRegression()
[0.5 0.5]
```

LinearRegression 拟合一个带有系数 $w = (w\_1, ..., w\_p)$ 的线性模型，使得数据集实际观测数据和预测数据（估计值）之间的残差平方和最小。其数学表达式为：

$$\min_{w} \|Xw - y\|_2^2$$

第 2 章

加载数据集

作为 Python 中经典的机器学习模块,sklearn 围绕着机器学习提供了很多可直接调用的机器学习算法以及很多经典的数据集。在 Scikit-Learn 应用中,使用 sklearn.datasets 加载数据集。在本章的内容中,将详细介绍使用 Scikit-Learn 实现数据集加载的知识,为读者步入本书后面知识的学习打下基础。

2.1 标准数据集 API

在 Scikit-Learn 中自带了一些小的标准数据集，不需要从外部网站下载任何文件，可以使用表 2-1 中的函数加载它们。

表 2-1 加载数据集函数　　　　　　　　　　　　　　　扫码观看本节视频讲解

函　　数	说　　明
load_boston(*[, return_X_y])	加载并返回波士顿房价数据集（回归）
load_iris(*[, return_X_y, as_frame])	加载并返回鸢尾花数据集（分类）
load_diabetes(*[, return_X_y, as_frame])	加载并返回糖尿病数据集（回归）
load_digits(*[, n_class, return_X_y, as_frame])	加载并返回手写数字数据集（分类）
load_linnerud(*[, return_X_y, as_frame])	加载并返回 physical Exercise linnerud 数据集
load_wine(*[, return_X_y, as_frame])	加载并返回红酒数据集（分类）
load_breast_cancer(*[, return_X_y, as_frame])	加载并返回威斯康星州乳腺癌数据集（分类）

2.1.1 波士顿房价数据集（适用于回归任务）

波士顿房价数据集包含了 506 处波士顿不同地理位置房产的房价数据（因变量），以及与之对应的包含房屋及房屋周围的详细信息（自变量），其中包含城镇犯罪率、一氧化氮浓度、住宅平均房间数、到中心区域的加权距离以及自住房平均房价等 13 个维度的数据，因此，波士顿房价数据集能够应用到回归问题上。这里使用 load_boston(return_X_y) 方法来导出数据，其中，参数 return_X_y 控制输出数据的结构，若设置为 True，则将因变量和自变量独立导出。

```
from sklearn import datasets

'''清空sklearn环境下所有数据'''
datasets.clear_data_home()

'''载入波士顿房价数据'''
X,y = datasets.load_boston(return_X_y=True)

'''获取自变量数据的形状'''
print(X.shape)

'''获取因变量数据的形状'''
print(y.shape)
```

执行后会输出：

```
(506, 13)
(506,)
```

2.1.2 威斯康星州乳腺癌数据集（适用于分类问题）

威斯康星州乳腺癌数据集包含了威斯康星州记录的 569 个乳腺癌病人的恶性／良性（1/0）类别型数据（训练目标），以及与之对应的 30 个维度的生理指标数据，因此这是一个非常标准的二类判别数据集。在这里使用 load_breast_cancer(return_X_y) 来导出数据。

```python
from sklearn import datasets

'''载入威斯康星州乳腺癌数据'''

X,y = datasets.load_breast_cancer(return_X_y=True)

'''获取自变量数据的形状'''

print(X.shape)

'''获取因变量数据的形状'''

print(y.shape)
```

执行后会输出：

```
In[21]: print(X.shape)
(569, 30)
In[22]: print(y.shape)
(569,)
```

2.1.3 糖尿病数据集（适用于回归任务）

糖尿病数据集主要包括 442 行数据、10 个属性值，分别是年龄（age）、性别（sex）、体质指数（body mass index）、平均血压（average blood pressure）、S1~S6 一年后疾病级数指标。Target 为一年后患疾病的定量指标，因此适用于回归任务。这里使用 load_diabetes(return_X_y) 来导出数据。

```python
from sklearn import datasets

'''载入糖尿病数据'''

X,y = datasets.load_diabetes(return_X_y=True)

'''获取自变量数据的形状'''

print(X.shape)

'''获取因变量数据的形状'''

print(y.shape)
```

执行后会输出：

```
(442, 10)
(442,)
```

2.1.4 手写数字数据集（适用于分类任务）

手写数字数据集是结构化数据的经典数据，共有 1797 个样本，每个样本有 64 个元素，对应到一个 8×8 像素点组成的矩阵，每一个值是其灰度值。我们都知道图片在计算机的底层实际是矩阵，每个位置对应一个像素点，有二值图、灰度图、1600 万色图等类型。在这个样本中对应的是灰度图，需要控制每一个像素的黑白浓淡，所以每个样本还原到矩阵后代表一个手写体数字，这与我们之前接触的数据有很大区别。在这里使用 load_digits(return_X_y) 来导出数据。

```
from sklearn import datasets
'''载入手写数字数据'''
data,target = datasets.load_digits(return_X_y=True)
print(data.shape)
print(target.shape)
```

执行后会输出：

```
(1797, 64)
(1797,)
```

下面是使用 matshow() 绘制矩阵形式的数据示意图的代码。

```
import matplotlib.pyplot as plt
import numpy as np

'''绘制数字0'''
num = np.array(data[0]).reshape((8,8))
plt.matshow(num)
print(target[0])

'''绘制数字5'''
num = np.array(data[5]).reshape((8,8))
plt.matshow(num)
print(target[5])

'''绘制数字9'''
num = np.array(data[9]).reshape((8,8))
plt.matshow(num)
print(target[9])
```

手写数字执行效果如图 2-1 所示。

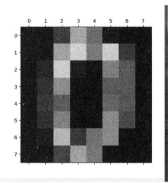 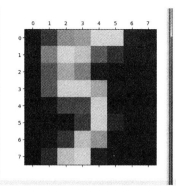 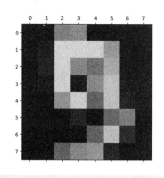

图 2-1 手写数字执行效果

2.1.5 Fisher 的鸢尾花数据集（适用于分类问题）

著名的统计学家 Fisher 在研究判别分析问题时收集了关于鸢尾花的一些数据，这是一个非常经典的数据集，datasets 中自然也带有这个数据集。这个数据集包含了 150 个鸢尾花样本，对应 3 种鸢尾花，各 50 个样本（target），以及它们各自对应的 4 种关于花外形的数据（自变量）。这里使用 load_iris(return_X_y) 来导出数据。

```
from sklearn import datasets

'''载入Fisher的鸢尾花数据'''

data,target = datasets.load_iris(return_X_y=True)

'''显示自变量的形状'''
print(data.shape)

'''显示训练目标的形状'''
print(target.shape)
```

执行后会输出：

```
(150, 4)
(150,)
```

2.1.6 红酒数据集（适用于分类问题）

这个数据集共 178 个样本，代表了红酒的三个档次（分别有 59、71、48 个样本），以及与之对应的 13 维的属性数据集，非常适合用来练习各种分类算法。在这里使用 load_wine(return_X_y) 来导出数据。

```
from sklearn import datasets

'''载入wine数据'''

data,target = datasets.load_wine(return_X_y=True)

'''显示自变量的形状'''
print(data.shape)

'''显示训练目标的形状'''
print(target.shape)
```

执行后会输出：

```
(178, 13)
(178,)
```

2.2 自定义数据集

本章前面介绍了几种 datasets 自带的经典数据集，但有时候我们需要自定义生成服从某些分布或者某些形状的数据集，而在 datasets 中就提供了这样的一些方法。

扫码观看本节视频讲解

2.2.1 生成聚类数据

在 Scikit-Learn 中，可以使用函数 datasets.make_blobs(n_samples, n_features,centers,cluster_std,center_box,shuffle,random_state) 生成各向同性的用于聚类的高斯点。其中，
- n_samples：控制随机样本点的个数；
- n_features：控制产生样本点的维度（对应n维正态分布）；
- centers：控制产生的聚类簇的个数。

例如下面的代码：

```
from sklearn import datasets
import matplotlib.pyplot as plt

X,y = datasets.make_blobs(n_samples=1000, n_features=2, centers=4, cluster_std=1.0,
center_box=(-10.0, 10.0), shuffle=True, random_state=None)

plt.scatter(X[:,0],X[:,1],c=y,s=8)
```

生成聚类数据，执行效果如图 2-2 所示。

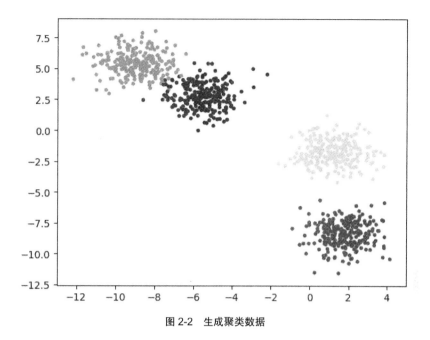

图 2-2 生成聚类数据

2.2.2 生成同心圆样本点

在 Scikit-Learn 中，可以使用函数 datasets.make_circles(n_samples,shuffle,noise,random_state,factor) 生成同心圆样本点。其中，
- n_samples：控制样本点总数；
- noise：控制属于同一个圈的样本点附加的漂移程度；
- factor：控制内外圈的接近程度，越大越接近，上限为1。

例如下面的代码：

```
from sklearn import datasets
import matplotlib.pyplot as plt

X,y = datasets.make_circles(n_samples=10000, shuffle=True, noise=0.04, random_state=None, factor=0.8)

plt.scatter(X[:,0],X[:,1],c=y,s=8)
```

生成同心圆样本点，执行效果如图 2-3 所示。

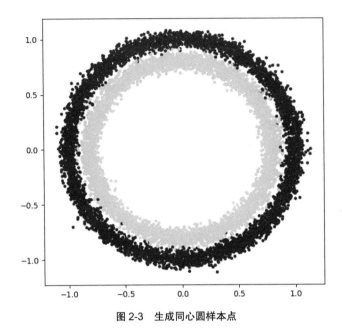

图 2-3 生成同心圆样本点

2.2.3 生成模拟分类数据集

在 Scikit-Learn 中，可以使用函数 datasets.make_classification(n_samples,n_features,n_informative,n_redundant,n_repeated,n_classes,n_clusters_per_class,weights,flip_y,class_sep,hypercube,shift,scale,shuffle,random_state) 生成模拟分类数据集。其中，

- n_samples：控制生成的样本点的个数；
- n_features：控制与类别有关的自变量的维数；
- n_classes：控制生成的分类数据类别的数量。

例如下面的代码：

```
from sklearn import datasets

X,y = datasets.make_classification(n_samples=100, n_features=20, n_informative=2, n_redundant=2, n_repeated=0, n_classes=2, n_clusters_per_class=2, weights=None, flip_y=0.01, class_sep=1.0, hypercube=True, shift=0.0, scale=1.0, shuffle=True, random_state=None)

print(X.shape)
print(y.shape)
set(y)
```

执行后会输出：

```
(100, 20)
(100,)
Out[2]: {0, 1}
```

2.2.4 生成太极型非凸集样本点

在 Scikit-Learn 中，可以使用函数 datasets.make_moons(n_samples,shuffle,noise,random_state) 生成太极型非凸集样本点。例如下面的代码：

```
from sklearn import datasets
import matplotlib.pyplot as plt

X,y = datasets.make_moons(n_samples=1000, shuffle=True, noise=0.05, random_state=None)

plt.scatter(X[:,0],X[:,1],c=y,s=8)
```

生成太极型非凸集样本点，执行效果如图 2-4 所示。

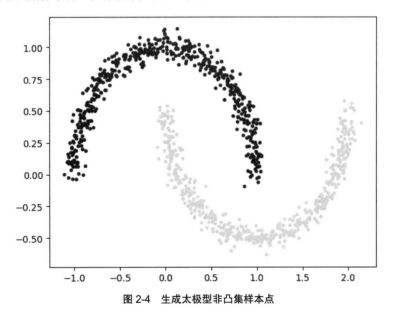

图 2-4 生成太极型非凸集样本点

第 3 章

监督学习

监督学习是从标记的训练数据来推断一个功能的机器学习任务。在监督学习中,每个实例都是由一个输入对象(通常为矢量)和一个期望的输出值(也称为监督信号)组成。监督学习算法是分析该训练数据并产生一个推断的功能,其可以用于映射出新的实例。在本章的内容中,将详细讲解使用 Scikit-Learn 实现监督学习的知识,为读者步入本书后面知识的学习打下基础。

3.1 广义线性模型

广义线性模型是线性模型的扩展,它通过联结函数建立响应变量的数学期望值与线性组合的预测变量之间的关系,其特点是不强行改变数据的自然度量,数据可以具有非线性和非恒定方差结构。

扫码观看本节视频讲解

下面是一组用于回归的方法,其中,目标值 y 是输入变量 x 的线性组合。在数学概念中,如果 $\hat{y}$ 是预测值,则:

$$\hat{y}(w,x) = w_0 + w_1 x_1 + \cdots + w_p x_p$$

在整个模块中,我们定义向量 $w = (w\_1,..., w\_p)$ 作为 coef_ ,定义 w_0 作为 intercept_ 。

3.1.1 普通最小二乘法

在 Scikit-Learn 应用中,模块 Linear Regression 用于拟合一个带有系数 $w = (w\_1, ..., w\_p)$ 的线性模型,使得数据集实际观测数据和预测数据(估计值)之间的残差平方和最小。其数学表达式为:

$$\min_{w} \|Xw - y\|_2^2$$

Linear Regression 会调用方法 fit() 拟合数组 X、y,并且将线性模型的系数 w 存储在其成员变量 coef_ 中。

然而,对于普通最小二乘的系数估计问题,其依赖于模型各项的相互独立性。当各项是相关的,且设计矩阵 X 的各列近似线性相关,那么设计矩阵会趋向于奇异矩阵,这会导致最小二乘估计对于随机误差非常敏感,产生很大的方差。例如,在没有实验设计的情况下收集到的数据,这种多重共线性(multicollinearity)的情况可能真的会出现。

请看下面的实例文件 linear01.py,使用第三方糖尿病数据集 diabetes 中的第一个特征说明二维图中的数据点。在本实例中使用线性回归绘制一条直线,以最小化数据集中观察到的响应与线性近似预测的响应之间的残差平方和进行绘制。

```
import matplotlib.pyplot as plt
import numpy as np
from sklearn import datasets, linear_model
from sklearn.metrics import mean_squared_error, r2_score

#加载糖尿病数据集
diabetes_X, diabetes_y = datasets.load_diabetes(return_X_y=True)

#只使用一个功能
diabetes_X = diabetes_X[:, np.newaxis, 2]

#将数据拆分为"训练/测试"集
diabetes_X_train = diabetes_X[:-20]
diabetes_X_test = diabetes_X[-20:]
```

```
#将目标划分为"训练/测试"集
diabetes_y_train = diabetes_y[:-20]
diabetes_y_test = diabetes_y[-20:]

#创建线性回归对象
regr = linear_model.LinearRegression()

#使用训练集训练模型
regr.fit(diabetes_X_train, diabetes_y_train)

#使用测试集进行预测
diabetes_y_pred = regr.predict(diabetes_X_test)

#系数
print('系数: \n', regr.coef_)
#均方误差
print('均方误差: %.2f'
      % mean_squared_error(diabetes_y_test, diabetes_y_pred))
#确定系数:如果是1则为完美预测
print('确定系数: %.2f'
      % r2_score(diabetes_y_test, diabetes_y_pred))

#输出绘图
plt.scatter(diabetes_X_test, diabetes_y_test,  color='black')
plt.plot(diabetes_X_test, diabetes_y_pred, color='blue', linewidth=3)

plt.xticks(())
plt.yticks(())

plt.show()
```

执行后会输出下面的结果。绘制可视化图表，如图3-1所示。

系数:
 [938.23786125]
均方误差: 2548.07
确定系数: 0.47

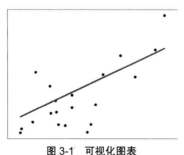

图3-1　可视化图表

3.1.2 岭回归

岭回归（Ridge Regression）是一种专门用于共线性数据分析的有偏估计回归方法，实质上是一种改良的最小二乘估计法，即通过放弃最小二乘法的无偏性，以损失部分信息、降低精度为代价获得回归系数更符合实际、更可靠的回归方法，它对病态数据的拟合要强于最小二乘法。

在 Scikit-Learn 应用中，使用模块 Ridge 实现岭回归功能。Ridge 回归通过对系数的大小施加惩罚来解决普通最小二乘法的一些问题。岭系数最小化的是带罚项的残差平方和：

$$\min_{w} \|Xw - y\|_2^2 + \alpha \|w\|_2^2$$

其中，$\alpha \geq 0$，是控制系数收缩量的复杂性参数：α 的值越大，收缩量越大，这样系数对共线性的鲁棒性也更强。与其他线性模型一样，Ridge 用方法 fit() 将模型系数 w 存储在其 coef_ 成员中，例如下面的代码：

```
>>> from sklearn import linear_model
>>> reg = linear_model.Ridge (alpha = .5)
>>> reg.fit ([[0, 0], [0, 0], [1, 1]], [0, .1, 1])
Ridge(alpha=0.5, copy_X=True, fit_intercept=True, max_iter=None,
 normalize=False, random_state=None, solver='auto', tol=0.001)
>>> reg.coef_
array([ 0.34545455,  0.34545455])
>>> reg.intercept_
0.13636...
```

请看下面的实例文件 linear02.py，其功能是绘制岭系数作为正则化的函数，可视化显示估计量系数中共线性的影响。

```python
import numpy as np
import matplotlib.pyplot as plt
from sklearn import linear_model

#X是10x10的希尔伯特矩阵
X = 1. / (np.arange(1, 11) + np.arange(0, 10)[:, np.newaxis])
y = np.ones(10)

#计算路径
n_alphas = 200
alphas = np.logspace(-10, -2, n_alphas)

coefs = []
for a in alphas:
    ridge = linear_model.Ridge(alpha=a, fit_intercept=False)
    ridge.fit(X, y)
    coefs.append(ridge.coef_)


#显示结果
ax = plt.gca()
```

```
ax.plot(alphas, coefs)
ax.set_xscale('log')
ax.set_xlim(ax.get_xlim()[::-1])  # reverse axis
plt.xlabel('alpha')
plt.ylabel('weights')
plt.title('作为正则化函数的岭系数')
plt.axis('tight')
plt.show()
```

在上述代码中，Ridge 回归是本例中使用的估计量，每种颜色代表系数向量的不同特征，这显示为正则化参数的函数。本实例还展示了将 Ridge 回归应用于高度病态矩阵的有用性，对于此类矩阵来说，即使目标变量的微小变化也会导致计算出的权重出现巨大差异。在这种情况下，设置某个正则化（alpha）以减少这种变化（噪声）很有用。

当 alpha 非常大时，正则化效应在平方损失函数中占主导地位，系数趋于零。在路径的末端，由于 alpha 趋于零且解趋于普通最小二乘法，系数表现出很大的振荡。在实践中，有必要以在两者之间保持平衡的方式调整 alpha。

3.1.3 Lasso 回归

在 Scikit-Learn 应用中，使用内置模块实现 Lasso 回归功能。Lasso 是估计稀疏系数的线性模型，这在一些情况下是有用的，因为它倾向于使用具有较少参数值的情况，有效地减少给定解决方案所依赖变量的数量。因此，Lasso 及其变体是压缩感知领域的基础。在一定条件下，它可以恢复一组非零权重的精确集。

在数学公式表达上，Lasso 由一个带有 e_1 先验的正则项的线性模型组成。其最小化的目标函数是：

$$\min_w \frac{1}{2n_{\text{samples}}} \|Xw - y\|_2^2 + \alpha \|w\|_1$$

Lasso 回归解决了加上罚项 $\alpha\|w\|_1$ 的最小二乘法的最小化问题，其中，α 是一个常数，$\|w\|_1$ 是参数向量的 e_1 – norm 范数（是对函数、向量和矩阵定义的一种度量形式）。例如下面的代码：

```
>>> from sklearn import linear_model
>>> reg = linear_model.Lasso(alpha=0.1)
>>> reg.fit([[0, 0], [1, 1]], [0, 1])
Lasso(alpha=0.1)
>>> reg.predict([[1, 1]])
array([0.8])
```

请看下面的实例文件 linear03.py，功能是实现稀疏信号的 Lasso 和弹性网。

```
import numpy as np
import matplotlib.pyplot as plt

from sklearn.metrics import r2_score
```

```python
# ############################################################
#生成一些稀疏数据以供使用
np.random.seed(42)

n_samples, n_features = 50, 100
X = np.random.randn(n_samples, n_features)

#递减系数,可视化的交替标志
idx = np.arange(n_features)
coef = (-1) ** idx * np.exp(-idx / 10)
coef[10:] = 0  #稀疏系数
y = np.dot(X, coef)

#增加噪声
y += 0.01 * np.random.normal(size=n_samples)

#分割、训练数据
n_samples = X.shape[0]
X_train, y_train = X[:n_samples // 2], y[:n_samples // 2]
X_test, y_test = X[n_samples // 2:], y[n_samples // 2:]

# ############################################################
#
from sklearn.linear_model import Lasso

alpha = 0.1
lasso = Lasso(alpha=alpha)

y_pred_lasso = lasso.fit(X_train, y_train).predict(X_test)
r2_score_lasso = r2_score(y_test, y_pred_lasso)
print(lasso)
print("r^2 on test data : %f" % r2_score_lasso)

# ############################################################
#弹力网
from sklearn.linear_model import ElasticNet

enet = ElasticNet(alpha=alpha, l1_ratio=0.7)

y_pred_enet = enet.fit(X_train, y_train).predict(X_test)
r2_score_enet = r2_score(y_test, y_pred_enet)
print(enet)
print("r^2 on test data : %f" % r2_score_enet)

m, s, _ = plt.stem(np.where(enet.coef_)[0], enet.coef_[enet.coef_ != 0],
                   markerfmt='x', label='Elastic net coefficients',
                   use_line_collection=True)
plt.setp([m, s], color="#2ca02c")
m, s, _ = plt.stem(np.where(lasso.coef_)[0], lasso.coef_[lasso.coef_ != 0],
```

```
            markerfmt='x', label='Lasso coefficients',
            use_line_collection=True)
plt.setp([m, s], color='#ff7f0e')
plt.stem(np.where(coef)[0], coef[coef != 0], label='true coefficients',
         markerfmt='bx', use_line_collection=True)

plt.legend(loc='best')
plt.title("Lasso $R^2$: %.3f, Elastic Net $R^2$: %.3f"
          % (r2_score_lasso, r2_score_enet))
plt.show()
```

Lasso 回归的执行效果如图 3-2 所示。

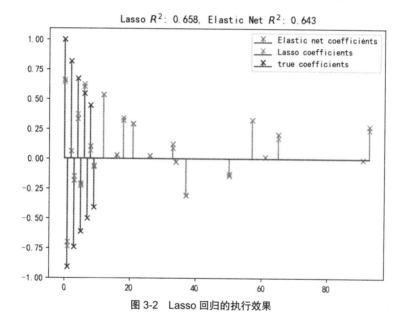

图 3-2 Lasso 回归的执行效果

3.2 线性判别分析和二次判别分析

线性判别分析（discriminant_analysis.Linear Discriminant Analysis，LDA）和二次判别分析（discriminant_analysis.Quadratic Discriminant Analysis，QDA）是两个经典的分类器，正如它们名字所描述的那样，它们分别代表了线性决策平面和二次决策平面。如图 3-3 所示，这些图像展示了线性判别分析以及二次判别分析的决策边界。其中，最后一行表明了线性判别分析只能学习线性边界，而二次判别分析则可以学习二次边界，因此它相对而言更加灵活。

扫码观看本节视频讲解

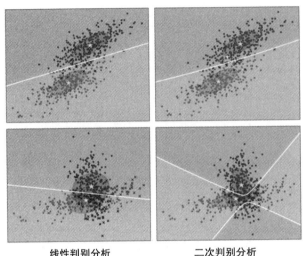

线性判别分析　　　　　　　　　二次判别分析

图 3-3　线性判别分析与二次判别分析对比

3.2.1　使用线性判别分析来降维

在 Scikit-Learn 应用中，线性判别分析通过把输入的数据投影到由最大化类之间分离的方向所组成的线性子空间，可以执行有监督降维。输出的维度必然会比原来的类别数量更少，因此线性判别分析总体而言是十分强大的降维方式。

关于维度的数量可以通过参数 n_components 来调节，值得注意的是，这个参数不会对方法 discriminant_analysis.LinearDiscriminantAnalysis.fit () 或者方法 discriminant_analysis.LinearDiscriminantAnalysis.predict () 产生影响。

请看下面的实例文件 linear04.py，功能是在 iris 数据集中对比 LDA 和 PCA 之间的降维差异。在 iris 数据集中保存了 3 种鸢尾花（Setosa、Versicolour 和 Virginica）信息，有 4 个属性：萼片长度、萼片宽度、花瓣长度和花瓣宽度。本实例将基于 iris 数据集分析应用于此数据的主成分（PCA），可以分析出造成数据差异最大的属性（主要成分或特征空间中的方向）组合。在这里，我们在两个第一主成分上绘制了不同的样本。

```
import matplotlib.pyplot as plt

from sklearn import datasets
from sklearn.decomposition import PCA
from sklearn.discriminant_analysis import LinearDiscriminantAnalysis

iris = datasets.load_iris()

X = iris.data
y = iris.target
target_names = iris.target_names
```

```
pca = PCA(n_components=2)
X_r = pca.fit(X).transform(X)

lda = LinearDiscriminantAnalysis(n_components=2)
X_r2 = lda.fit(X, y).transform(X)

#为每个组成部分计算差异百分比
print('explained variance ratio (first two components): %s'
      % str(pca.explained_variance_ratio_))

plt.figure()
colors = ['navy', 'turquoise', 'darkorange']
lw = 2

for color, i, target_name in zip(colors, [0, 1, 2], target_names):
    plt.scatter(X_r[y == i, 0], X_r[y == i, 1], color=color, alpha=.8, lw=lw,
                label=target_name)
plt.legend(loc='best', shadow=False, scatterpoints=1)
plt.title('PCA of IRIS dataset')

plt.figure()
for color, i, target_name in zip(colors, [0, 1, 2], target_names):
    plt.scatter(X_r2[y == i, 0], X_r2[y == i, 1], alpha=.8, color=color,
                label=target_name)
plt.legend(loc='best', shadow=False, scatterpoints=1)
plt.title('LDA of IRIS dataset')

plt.show()
```

运行上述代码，LDA 会试图识别出类别之间差异最大的属性。尤其是与 PCA 相比，LDA 是使用已知类别标签的受监督方法。执行后会输出下面的结果，并绘制如图 3-4 所示的可视化对比图。

```
explained variance ratio (first two components): [0.92461872 0.05306648]
```

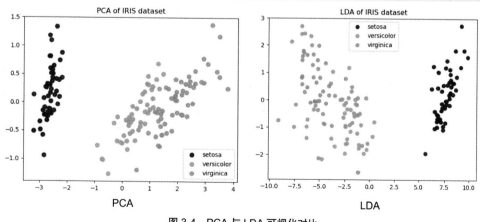

图 3-4　PCA 与 LDA 可视化对比

3.2.2 LDA 和 QDA 分类器的数学公式

LDA 和 QDA 都是源于简单的概率模型，这些模型对于每一个类别 k 的相关分布 $P(X|y=k)$ 都可以通过贝叶斯定理获得。

$$P(y=k|X) = \frac{P(X|y=k)P(y=k)}{P(X)} = \frac{P(X|y=k)P(y=k)}{\sum_l P(X|y=l) \cdot P(y=l)}$$

我们选择最大化条件概率的类别 k，更具体地说，对于 LDA 以及 QDA，$P(X|y)$ 被建模呈密度多变量高斯分布：

$$P(X|y=k) = \frac{1}{(2\pi)^n |\Sigma_k|^{1/2}} \exp\left[-\frac{1}{2}(X-\mu_k)^t \Sigma_k^{-1}(X-\mu_k)\right]$$

为了把该模型作为分类器使用，我们只需要从训练数据中估计出类的先验概率 $P(y=k)$（通过每个类 k 的实例的比例得到）、类别均值 μ_k（通过经验样本的类别均值得到）以及协方差矩阵（通过经验样本的类别协方差或者正则化的估计器 estimator 得到）。

为了理解 LDA 在降维上的应用，从上面解释的 LDA 分类规则的几何重构开始是十分有用的。我们用 K 表示目标类别的总数。 由于在 LDA 中我们假设所有类别都有相同估计的协方差 Σ，所以可重新调节数据从而让协方差相同。

$$X^* = D^{-1/2}U^t X \text{ with } \Sigma = UDU^t$$

在缩放之后对数据点进行分类，相当于找到与欧几里得距离中的数据点最接近的估计类别均值。但是它也可以在投影到 $K-1$ 个由所有类中的所有的 μ_k^* 生成的仿射子空间 H_K 之后完成，这表明在 LDA 分类器中存在一个利用线性投影到 $K-1$ 个维度空间的降维工具。

通过投影到线性子空间 H_L 上，可以进一步将维数减少到一个选定的 L，从而使投影后的 μ_k^* 的方差最大化（实际上，为了实现转换类均值 μ_k^*，我们正在做一种形式的 PCA）。 这里的 L 对应于在方法 discriminant_analysis.LinearDiscriminantAnalysis.transform() 中使用的参数 n_components。

3.2.3 收缩

收缩 (Shrinkage) 是一种在训练样本数量相比特征而言很小的情况下，可以提升的协方差矩阵预测（准确性）的工具。在这种情况下，经验样本协方差是一个很差的预测器。收缩 LDA 可以通过设置类 discriminant_analysis.LinearDiscriminantAnalysis 的参数 shrinkage 为 "auto" 来实现。

我们可以手动将收缩参数（shrinkage parameter）的值设置为 0~1，其中，0 对应着没有收缩（这意味着经验协方差矩阵将会被使用），而 1 则对应着完全使用收缩（意味着方差的对角矩阵将被当作协方差矩阵的估计）。设置该参数在两个极端值之间会估计一个（特定的）协方差矩阵的收缩形式。

请看下面的实例文件 linear05.py，功能是实现用于分类的正态、Ledoit-Wolf 和 Oracle Shrinkage Approximating (OSA) 线性判断分析。本实例演示了使用 Ledoit-Wolf 和 OSA 协方差估计器改进分类的方法。

```
import numpy as np
import matplotlib.pyplot as plt
```

```python
from sklearn.datasets import make_blobs
from sklearn.discriminant_analysis import LinearDiscriminantAnalysis
from sklearn.covariance import OAS


n_train = 20    #训练样本
n_test = 200    #测试样本
n_averages = 50   #重复分类的频率
n_features_max = 75   #最大功能数
step = 4   #计算的步长


def generate_data(n_samples, n_features):
    """生成带有噪声特征的随机斑点数据,这将返回一个具有(n_samples, n_features)
    输入数据的驻足以及n_samples目标标签数组.
    只有一个特征包含鉴别信息,其他特征只包含噪声.
    """
    X, y = make_blobs(n_samples=n_samples, n_features=1, centers=[[-2], [2]])

    #添加非歧视性特征
    if n_features > 1:
        X = np.hstack([X, np.random.randn(n_samples, n_features - 1)])
    return X, y


acc_clf1, acc_clf2, acc_clf3 = [], [], []
n_features_range = range(1, n_features_max + 1, step)
for n_features in n_features_range:
    score_clf1, score_clf2, score_clf3 = 0, 0, 0
    for _ in range(n_averages):
        X, y = generate_data(n_train, n_features)

        clf1 = LinearDiscriminantAnalysis(solver='lsqr',
                                          shrinkage='auto').fit(X, y)
        clf2 = LinearDiscriminantAnalysis(solver='lsqr',
                                          shrinkage=None).fit(X, y)
        oa = OAS(store_precision=False, assume_centered=False)
        clf3 = LinearDiscriminantAnalysis(solver='lsqr',
                                          covariance_estimator=oa).fit(X, y)

        X, y = generate_data(n_test, n_features)
        score_clf1 += clf1.score(X, y)
        score_clf2 += clf2.score(X, y)
        score_clf3 += clf3.score(X, y)

    acc_clf1.append(score_clf1 / n_averages)
    acc_clf2.append(score_clf2 / n_averages)
    acc_clf3.append(score_clf3 / n_averages)
```

```
features_samples_ratio = np.array(n_features_range) / n_train

plt.plot(features_samples_ratio, acc_clf1, linewidth=2,
        label="Linear Discriminant Analysis with Ledoit Wolf", color='navy')
plt.plot(features_samples_ratio, acc_clf2, linewidth=2,
        label="Linear Discriminant Analysis", color='gold')
plt.plot(features_samples_ratio, acc_clf3, linewidth=2,
        label="Linear Discriminant Analysis with OAS", color='red')

plt.xlabel('n_features / n_samples')
plt.ylabel('Classification accuracy')

plt.legend(loc=3, prop={'size': 12})
plt.suptitle('Linear Discriminant Analysis vs. ' + '\n'
            + 'Shrinkage Linear Discriminant Analysis vs. ' + '\n'
            + 'OAS Linear Discriminant Analysis (1 discriminative feature)')
plt.show()
```

linear05.py 的执行效果如图 3-5 所示。

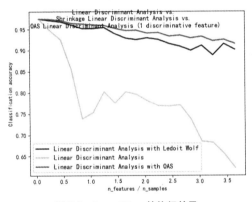

图 3-5　linear05.py 的执行效果

3.3　内核岭回归

内核岭回归（kernel ridge regression，KRR）由使用内核方法的岭回归（使用 L2 正则化的最小二乘法）组成。因此，它所拟合到的在空间中不同的线性函数是由不同的内核和数据所导致的。对于非线性的内核，它与原始空间中的非线性函数相对应。

由 KernelRidge 学习的模型的形式与支持向量回归（support vector regression，SVR）是一样的，但是它们使用不同的损失函数：内核岭回归（KRR）使用 squared error loss（平方误差损失函数），而支持向量回归（SVR）使用 ε-insensitive loss（ε-不敏感损失），两者都使用 L2 regularization（L2 正则化）。与 SVR 相反，拟合 KernelRidge 可以以 closed-

扫码观看本节视频讲解

form（封闭形式）完成，对于中型数据集通常更快。另外，学习的模型是非稀疏的，因此比 SVR 慢，关于预测时间，SVR 学习了 :math:epsilon > 0 的稀疏模型。

图 3-6 比较了人造数据集上的 KernelRidge 和 SVR 的区别，它由一个正弦目标函数和每 5 个数据点产生一个强噪声组成。图 3-6 分别绘制了由 KernelRidge 和 SVR 学习到的回归曲线。两者都使用网格搜索优化了 RBF 内核的 complexity/regularization（复杂性 / 正则化）和 bandwidth（带宽）。它们的 learned functions（学习函数）非常相似，但是，拟合 KernelRidge 大约比拟合 SVR 快 7 倍 [都使用 grid-search（网格搜索）]。然而，由于 SVR 只学习了一个稀疏模型，所以 SVR 预测 10 万个目标值比使用 KernelRidge 快 3 倍以上。SVR 只使用了 30% 的数据点作为支撑向量。

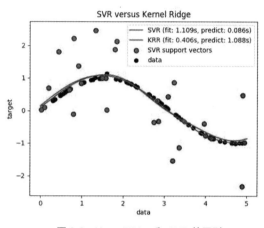

图 3-6　KernelRidge 和 SVR 的区别

在曲线下方显示了不同大小训练集的 KernelRidge 和 SVR 的拟合（fitting）和预测（prediction）时间。对于中型训练集（小于 1000 个样本），拟合 KernelRidge 比 SVR 快；然而，对于更大的训练集，SVR 通常更好。关于预测时间，由于学习的稀疏解，SVR 对于所有不同大小的训练集都比 KernelRidge 快。注意，稀疏度和预测时间取决于 SVR 的参数 ε 和 C，ε = 0 将对应于密集模型。

3.4　支持向量机

支持向量机（SVMs）可用于监督学习算法中的分类、回归和异常检测等。支持向量机的优势在于：
- 在高维空间中非常高效；
- 即使在数据维度比样本数量大的情况下仍然有效；
- 在决策函数（称为支持向量）中使用训练集的子集，因此它也是高效利用内存的。

扫码观看本节视频讲解

在 Scikit-Learn 中提供了 dense（numpy.ndarray，可以通过 numpy.asarray 进行转换）和 sparse（任何 scipy.sparse）实现支持向量机功能。要使用支持向量机来对 sparse 数据进行预测，它必须已经拟合这样的数据。

3.4.1 分类

在 Scikit-Learn 应用中，SVC、NuSVC 和 LinearSVC 能在数据集中实现多元分类功能。SVC 和 NuSVC 是相似的方法，但是支持 SVC 向量分类，接受稍许不同的参数设置并且有不同的数学方程。另外，LinearSVC 是另一个实现线性核函数的支持向量分类。LinearSVC 不接受关键词 kernel，因为它被假设为线性的，也缺少一些 SVC 和 NuSVC 的成员（members），比如 support_。

和其他分类器一样，SVC、NuSVC 和 LinearSVC 将两个数组作为输入，其中，[n_samples, n_features] 大小的数组 x 作为训练样本，[n_samples] 大小的数组 y 作为类别标签（字符串或整数）。例如下面的代码中，使用 SVC 对设置的数据 x 和 y 进行了拟合处理：

```
>>> from sklearn import svm
>>> X = [[0, 0], [1, 1]]
>>> y = [0, 1]
>>> clf = svm.SVC()
>>> clf.fit(X, y)
SVC(C=1.0, cache_size=200, class_weight=None, coef0=0.0,
    decision_function_shape='ovr', degree=3, gamma='auto', kernel='rbf',
    max_iter=-1, probability=False, random_state=None, shrinking=True,
    tol=0.001, verbose=False)
```

在拟合后，可以使用这个模型预测新的值：

```
>>> clf.predict([[2., 2.]])
array([1])
```

SVMs 决策函数取决于训练集的一些子集，称作支持向量。这些支持向量的部分特性可以在 support_vectors_、support_ 和 n_support_ 成员中找到。例如下面的代码：

```
>>> #获得支持向量
>>> clf.support_vectors_
array([[ 0.,  0.],
       [ 1.,  1.]])
>>> #获得支持向量的索引
>>> clf.support_
array([0, 1]...)
>>> #为每一个类别获得支持向量的数量
>>> clf.n_support_
array([1, 1]...)
```

请看下面的实例文件 linear06.py，功能是使用具有线性核的支持向量机分类器在两类可分离数据集中绘制最大边距分离超平面。

```
import numpy as np
import matplotlib.pyplot as plt
from sklearn import svm
from sklearn.datasets import make_blobs
```

```
#创造40个可分离的点
X, y = make_blobs(n_samples=40, centers=2, random_state=6)

#适合模型,不要为了说明而正则化
clf = svm.SVC(kernel='linear', C=1000)
clf.fit(X, y)

plt.scatter(X[:, 0], X[:, 1], c=y, s=30, cmap=plt.cm.Paired)

#绘制决策函数
ax = plt.gca()
xlim = ax.get_xlim()
ylim = ax.get_ylim()

#创建网格以评估模型
xx = np.linspace(xlim[0], xlim[1], 30)
yy = np.linspace(ylim[0], ylim[1], 30)
YY, XX = np.meshgrid(yy, xx)
xy = np.vstack([XX.ravel(), YY.ravel()]).T
Z = clf.decision_function(xy).reshape(XX.shape)

#绘图决策边界和边距
ax.contour(XX, YY, Z, colors='k', levels=[-1, 0, 1], alpha=0.5,
           linestyles=['--', '-', '--'])
#绘制支持向量
ax.scatter(clf.support_vectors_[:, 0], clf.support_vectors_[:, 1], s=100,
           linewidth=1, facecolors='none', edgecolors='k')
plt.show()
```

linear06.py 的执行效果如图 3-7 所示。

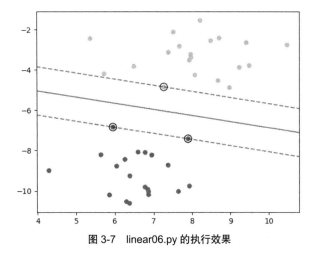

图 3-7　linear06.py 的执行效果

3.4.2 回归

支持向量分类的方法可以被扩展用作解决回归问题,这个方法被称为支持向量回归。支持向量分类生成的模型(如前文描述)只依赖于训练集的子集,因为构建模型的 cost function 不在乎边缘之外的训练点。类似地,支持向量回归生成的模型只依赖于训练集的子集,因为构建模型的 cost function 忽略任何接近于模型预测的训练数据。

支持向量分类有三种不同的实现形式:SVR、NuSVR 和 LinearSVR。在只考虑线性核的情况下,LinearSVR 比 SVR 提供了一个更快的实现形式,然而比起 SVR 和 LinearSVR,NuSVR 能够实现一个稍微不同的构思(formulation)。

方法 fit() 会调用参数向量 x、y,y 是浮点数,而不是整型数。例如下面的代码:

```
>>> from sklearn import svm
>>> X = [[0, 0], [2, 2]]
>>> y = [0.5, 2.5]
>>> clf = svm.SVR()
>>> clf.fit(X, y)
SVR(C=1.0, cache_size=200, coef0=0.0, degree=3, epsilon=0.1, gamma='auto',
 kernel='rbf', max_iter=-1, shrinking=True, tol=0.001, verbose=False)
>>> clf.predict([[1, 1]])
array([ 1.5])
```

请看下面的实例文件 linear07.py,功能是使用线性和非线性内核的支持向量回归。

```
import numpy as np
from sklearn.svm import SVR
import matplotlib.pyplot as plt

# ###############################################################
#生成示例数据
X = np.sort(5 * np.random.rand(40, 1), axis=0)
y = np.sin(X).ravel()

# ###############################################################
#向目标添加噪波
y[::5] += 3 * (0.5 - np.random.rand(8))

# ###############################################################
#拟合回归模型
svr_rbf = SVR(kernel='rbf', C=100, gamma=0.1, epsilon=.1)
svr_lin = SVR(kernel='linear', C=100, gamma='auto')
svr_poly = SVR(kernel='poly', C=100, gamma='auto', degree=3, epsilon=.1,
               coef0=1)

# ###############################################################
#查看结果
lw = 2
```

```python
svrs = [svr_rbf, svr_lin, svr_poly]
kernel_label = ['RBF', 'Linear', 'Polynomial']
model_color = ['m', 'c', 'g']

fig, axes = plt.subplots(nrows=1, ncols=3, figsize=(15, 10), sharey=True)
for ix, svr in enumerate(svrs):
    axes[ix].plot(X, svr.fit(X, y).predict(X), color=model_color[ix], lw=lw,
                  label='{} model'.format(kernel_label[ix]))
    axes[ix].scatter(X[svr.support_], y[svr.support_], facecolor="none",
                     edgecolor=model_color[ix], s=50,
                     label='{} support vectors'.format(kernel_label[ix]))
    axes[ix].scatter(X[np.setdiff1d(np.arange(len(X)), svr.support_)],
                     y[np.setdiff1d(np.arange(len(X)), svr.support_)],
                     facecolor="none", edgecolor="k", s=50,
                     label='other training data')
    axes[ix].legend(loc='upper center', bbox_to_anchor=(0.5, 1.1),
                    ncol=1, fancybox=True, shadow=True)

fig.text(0.5, 0.04, 'data', ha='center', va='center')
fig.text(0.06, 0.5, 'target', ha='center', va='center', rotation='vertical')
fig.suptitle("Support Vector Regression", fontsize=14)
plt.show()
```

linear07.py 的执行效果如图 3-8 所示。

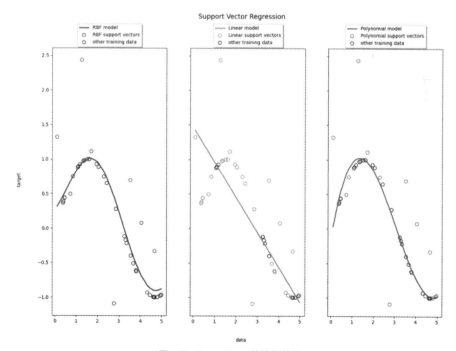

图 3-8　linear07.py 的执行效果

3.4.3 密度估计和异常检测

SVM 也可用于异常（novelty）检测，即给予一个样例集会检测这个样例集的 soft boundary（柔性界线），以便给新的数据点分类，看它是否属于这个样例集，生成的类被称为 OneClassSVM（异常点检测算法）。在这种情况下，因为它属于非监督学习的一类，所以没有类标签，方法 fit() 只会考虑输入数组 X。

请看下面的实例文件 linear08.py，功能是使用一类 SVM 实现新颖性检测功能。SVM 是一种无监督算法，学习用于新颖性检测的决策函数：将新数据分类为与训练集相似或不同。

```python
import numpy as np
import matplotlib.pyplot as plt
import matplotlib.font_manager
from sklearn import svm

xx, yy = np.meshgrid(np.linspace(-5, 5, 500), np.linspace(-5, 5, 500))
#生成训练数据
X = 0.3 * np.random.randn(100, 2)
X_train = np.r_[X + 2, X - 2]
#产生一些有规律的新的观察结果
X = 0.3 * np.random.randn(20, 2)
X_test = np.r_[X + 2, X - 2]
#产生一些不正常的新的观察结果
X_outliers = np.random.uniform(low=-4, high=4, size=(20, 2))

#fit模型
clf = svm.OneClassSVM(nu=0.1, kernel="rbf", gamma=0.1)
clf.fit(X_train)
y_pred_train = clf.predict(X_train)
y_pred_test = clf.predict(X_test)
y_pred_outliers = clf.predict(X_outliers)
n_error_train = y_pred_train[y_pred_train == -1].size
n_error_test = y_pred_test[y_pred_test == -1].size
n_error_outliers = y_pred_outliers[y_pred_outliers == 1].size

#绘制线、点和到平面最近的向量
Z = clf.decision_function(np.c_[xx.ravel(), yy.ravel()])
Z = Z.reshape(xx.shape)

plt.title("Novelty Detection")
plt.contourf(xx, yy, Z, levels=np.linspace(Z.min(), 0, 7), cmap=plt.cm.PuBu)
a = plt.contour(xx, yy, Z, levels=[0], linewidths=2, colors='darkred')
plt.contourf(xx, yy, Z, levels=[0, Z.max()], colors='palevioletred')

s = 40
b1 = plt.scatter(X_train[:, 0], X_train[:, 1], c='white', s=s, edgecolors='k')
b2 = plt.scatter(X_test[:, 0], X_test[:, 1], c='blueviolet', s=s,
                 edgecolors='k')
c = plt.scatter(X_outliers[:, 0], X_outliers[:, 1], c='gold', s=s,
```

```
            edgecolors='k')
plt.axis('tight')
plt.xlim((-5, 5))
plt.ylim((-5, 5))
plt.legend([a.collections[0], b1, b2, c],
           ["learned frontier", "training observations",
            "new regular observations", "new abnormal observations"],
           loc="upper left",
           prop=matplotlib.font_manager.FontProperties(size=11))
plt.xlabel(
    "error train: %d/200 ; errors novel regular: %d/40 ; "
    "errors novel abnormal: %d/40"
    % (n_error_train, n_error_test, n_error_outliers))
plt.show()
```

linear08.py 的执行效果如图 3-9 所示。

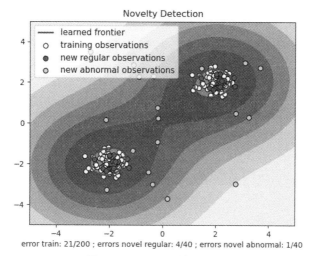

图 3-9　linear08.py 的执行效果

3.5　随机梯度下降

扫码观看本节视频讲解

　　随机梯度下降（SGD）是一种既简单又非常高效的方法，主要用于凸函数下线性分类器的判别式学习，例如（线性）支持向量机和 Logistic 回归。尽管 SGD 在机器学习社区已经存在了很长时间，但是最近在大规模学习（large-scale learning）方面 SGD 才获得了相当大的关注。SGD 已成功应用于文本分类和自然语言处理中经常遇到的大规模和稀疏的机器学习问题。对于稀疏数据来说，本模块的分类器可以轻易地处理超过 10^5 的训练样本和超过 10^5 的特征。

3.5.1 分类

在拟合模型前，需要确保重新排列（打乱）了训练数据，或者在每次迭代后用 shuffle=True 来打乱。类 SGDClassifier 实现了一个简单的随机梯度下降学习过程，支持不同的损失函数（loss functions）和分类处罚（penalties for classification）。作为另一个分类器（classifier），拟合 SGD 需要两个数组（array），其中一个是将训练样本的 size 保存为 [n_samples, n_features] 的数组 X，另一个是将训练样本目标值（类标签）的 size 保存为[n_samples] 的数组 Y。例如，下面的代码使用SGDClassifier对数据进行分类和拟合处理。

```
>>> from sklearn.linear_model import SGDClassifier
>>> X = [[0., 0.], [1., 1.]]
>>> y = [0, 1]
>>> clf = SGDClassifier(loss="hinge", penalty="l2")
>>> clf.fit(X, y)
SGDClassifier(alpha=0.0001, average=False, class_weight=None, epsilon=0.1,
 eta0=0.0, fit_intercept=True, l1_ratio=0.15,
 learning_rate='optimal', loss='hinge', max_iter=5, n_iter=None,
 n_jobs=1, penalty='l2', power_t=0.5, random_state=None,
 shuffle=True, tol=None, verbose=0, warm_start=False)
```

拟合后可以用该模型来预测新值：

```
>>> clf.predict([[2., 2.]])
array([1])
```

SGD 通过训练数据来拟合一个线性模型，成员 coef_ 用于保存模型参数：

```
>>> clf.coef_
array([[ 9.9...,  9.9...]])
```

成员 intercept_ 用于保存 intercept [截距，又称作偏移（offset）或偏差（bias）]：

```
>>> clf.intercept_
array([-9.9...])
```

使用 SGDClassifier.decision_function 来获得到此超平面的符号距离（signed distance）：

```
>>> clf.decision_function([[2., 2.]])
array([ 29.6...])
```

具体的 loss function 可以通过 loss 参数来设置，SGDClassifier 支持如下 loss functions。
- loss="hinge"：（软-间隔）线性支持向量机；
- loss="modified_huber"：平滑的 hinge 损失；
- loss="log"：logistic 回归；
- 所有的回归损失。

其中，前两个 loss functions 是懒惰的，如果一个例子违反了边界约束（margin constraint），它们仅更新模型的参数，这使得训练非常有效率，即使使用了 L2 惩罚（L2 penalty），我们仍然可以得到稀疏的模型结果。

使用 loss="log" 或者 loss="modified_huber" 来启用 predict_proba 方法，其给出每个样本 x 的概率估计 $P(y|x)$ 的一个向量：

```
>>> clf = SGDClassifier(loss="log").fit(X, y)
>>> clf.predict_proba([[1., 1.]])
array([[ 0.00...,  0.99...]])
```

请看下面的实例文件 linear09.py，功能是使用 SGD 训练的线性支持向量机分类器在二类可分离数据集中绘制最大间隔分离超平面。

```python
#创建50个可分离点
X, Y = make_blobs(n_samples=50, centers=2, random_state=0, cluster_std=0.60)

#fit模型
clf = SGDClassifier(loss="hinge", alpha=0.01, max_iter=200)

clf.fit(X, Y)

#绘制线、点和到平面最近的向量
xx = np.linspace(-1, 5, 10)
yy = np.linspace(-1, 5, 10)

X1, X2 = np.meshgrid(xx, yy)
Z = np.empty(X1.shape)
for (i, j), val in np.ndenumerate(X1):
    x1 = val
    x2 = X2[i, j]
    p = clf.decision_function([[x1, x2]])
    Z[i, j] = p[0]
levels = [-1.0, 0.0, 1.0]
linestyles = ['dashed', 'solid', 'dashed']
colors = 'k'
plt.contour(X1, X2, Z, levels, colors=colors, linestyles=linestyles)
plt.scatter(X[:, 0], X[:, 1], c=Y, cmap=plt.cm.Paired,
            edgecolor='black', s=20)

plt.axis('tight')
plt.show()
```

linear09.py 的执行效果如图 3-10 所示。

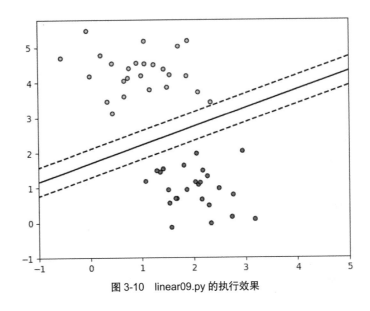

图 3-10　linear09.py 的执行效果

3.5.2　回归

SGDRegressor 类实现了一个简单的随机梯度下降学习例程，它支持用不同的损失函数和惩罚来拟合线性回归模型。SGDRegressor 非常适用于有大量训练样本（>10.000）的回归问题，对于其他问题，建议使用 Ridge、Lasso 或 ElasticNet。具体的损失函数可以通过 loss 参数设置，SGDRegressor 支持如下所示的损失函数。

- loss="squared_loss"：普通最小二乘法；
- loss="huber"：Huber回归；
- loss="epsilon_insensitive"：线性支持向量回归。

huber 和 epsilon-insensitive 损失函数可用于鲁棒回归（robust regression）。不敏感区域的宽度必须通过参数 epsilon 来设定，这个参数取决于目标变量的规模。

SGDRegressor 支持平均随机梯度下降（ASGD）作为 SGDClassifier，均值化可以通过设置 average=True 来启用。

对于利用了平方损失（squared loss）和 L2 惩罚（L2 penalty）的回归，在 Ridge 中提供了另一个采取平均策略（averaging strategy）的 SGD 变体，其使用了随机平均梯度（SAG）算法。

3.5.3　稀疏数据的随机梯度下降

由于在截距部分收敛学习效率方面的差异，稀疏实现与密集实现相比产生的结果略有不同。在 scipy .sparse 支持的格式中，任意矩阵都有对稀疏数据的内置支持方法。但是，为了获得最高的效率，请使用在 scipy.sparse.csr_matrix 中定义的 CSR 矩阵格式。

请看下面的实例文件 linear10.py，功能是展示使用词袋模型按主题对文档进行分类的过程。本实例使用 scipy.sparse 矩阵来存储特征，并演示了可以有效处理稀疏矩阵的各种分类器。实例文件 linear10.py 的具体实现流程如下。

（1）本实例使用的数据集由 20 个新闻组数据集组成，将自动下载这个数据集并缓存。代码如下：

```python
if opts.all_categories:
    categories = None
else:
    categories = [
        'alt.atheism',
        'talk.religion.misc',
        'comp.graphics',
        'sci.space',
    ]

if opts.filtered:
    remove = ('headers', 'footers', 'quotes')
else:
    remove = ()

print("Loading 20 newsgroups dataset for categories:")
print(categories if categories else "all")

data_train = fetch_20newsgroups(subset='train', categories=categories,
                                shuffle=True, random_state=42,
                                remove=remove)

data_test = fetch_20newsgroups(subset='test', categories=categories,
                               shuffle=True, random_state=42,
                               remove=remove)
print('data loaded')

#target_names中标签的顺序可以不同于categories
target_names = data_train.target_names


def size_mb(docs):
    return sum(len(s.encode('utf-8')) for s in docs) / 1e6


data_train_size_mb = size_mb(data_train.data)
data_test_size_mb = size_mb(data_test.data)

print("%d documents - %0.3fMB (training set)" % (
    len(data_train.data), data_train_size_mb))
print("%d documents - %0.3fMB (test set)" % (
    len(data_test.data), data_test_size_mb))
```

```python
print("%d categories" % len(target_names))
print()

#拆分训练集和测试集
y_train, y_test = data_train.target, data_test.target

print("Extracting features from the training data using a sparse vectorizer")
t0 = time()
if opts.use_hashing:
    vectorizer = HashingVectorizer(stop_words='english', alternate_sign=False,
                                   n_features=opts.n_features)
    X_train = vectorizer.transform(data_train.data)
else:
    vectorizer = TfidfVectorizer(sublinear_tf=True, max_df=0.5,
                                 stop_words='english')
    X_train = vectorizer.fit_transform(data_train.data)
duration = time() - t0
print("done in %fs at %0.3fMB/s" % (duration, data_train_size_mb / duration))
print("n_samples: %d, n_features: %d" % X_train.shape)
print()

print("Extracting features from the test data using the same vectorizer")
t0 = time()
X_test = vectorizer.transform(data_test.data)
duration = time() - t0
print("done in %fs at %0.3fMB/s" % (duration, data_test_size_mb / duration))
print("n_samples: %d, n_features: %d" % X_test.shape)
print()

#从整数特征名称到原始标记字符串的映射
if opts.use_hashing:
    feature_names = None
else:
    feature_names = vectorizer.get_feature_names()

if opts.select_chi2:
    print("Extracting %d best features by a chi-squared test" %
          opts.select_chi2)
    t0 = time()
    ch2 = SelectKBest(chi2, k=opts.select_chi2)
    X_train = ch2.fit_transform(X_train, y_train)
    X_test = ch2.transform(X_test)
    if feature_names:
        #保留选定的要素名称
        feature_names = [feature_names[i] for i
                         in ch2.get_support(indices=True)]
    print("done in %fs" % (time() - t0))
    print()
```

```python
if feature_names:
    feature_names = np.asarray(feature_names)

def trim(s):
    """格式化处理字符串，以适合的宽度显示结果(假设显示80列)"""
    return s if len(s) <= 80 else s[:77] + "..."
```

通过上述代码，从新闻组数据集加载数据，该数据集包含关于 20 个主题的大约 18000 个新闻组帖子，分为两个子集：一个用于训练或开发，另一个用于测试或性能评估。

（2）用 15 个不同的分类模型训练和测试数据集，并获得每个模型的性能结果。代码如下：

```python
def benchmark(clf):
    print('_' * 80)
    print("Training: ")
    print(clf)
    t0 = time()
    clf.fit(X_train, y_train)
    train_time = time() - t0
    print("train time: %0.3fs" % train_time)

    t0 = time()
    pred = clf.predict(X_test)
    test_time = time() - t0
    print("test time:  %0.3fs" % test_time)

    score = metrics.accuracy_score(y_test, pred)
    print("accuracy:   %0.3f" % score)

    if hasattr(clf, 'coef_'):
        print("dimensionality: %d" % clf.coef_.shape[1])
        print("density: %f" % density(clf.coef_))

        if opts.print_top10 and feature_names is not None:
            print("top 10 keywords per class:")
            for i, label in enumerate(target_names):
                top10 = np.argsort(clf.coef_[i])[-10:]
                print(trim("%s: %s" % (label, " ".join(feature_names[top10]))))
        print()

    if opts.print_report:
        print("classification report:")
        print(metrics.classification_report(y_test, pred,
                                            target_names=target_names))

    if opts.print_cm:
        print("confusion matrix:")
        print(metrics.confusion_matrix(y_test, pred))
```

```python
        print()
        clf_descr = str(clf).split('(')[0]
        return clf_descr, score, train_time, test_time


results = []
for clf, name in (
        (RidgeClassifier(tol=1e-2, solver="sag"), "Ridge Classifier"),
        (Perceptron(max_iter=50), "Perceptron"),
        (PassiveAggressiveClassifier(max_iter=50),
         "Passive-Aggressive"),
        (KNeighborsClassifier(n_neighbors=10), "kNN"),
        (RandomForestClassifier(), "Random forest")):
    print('=' * 80)
    print(name)
    results.append(benchmark(clf))

for penalty in ["l2", "l1"]:
    print('=' * 80)
    print("%s penalty" % penalty.upper())
    #训练Liblinear模型
    results.append(benchmark(LinearSVC(penalty=penalty, dual=False,
                                       tol=1e-3)))

    #训练 SGD 模型
    results.append(benchmark(SGDClassifier(alpha=.0001, max_iter=50,
                                           penalty=penalty)))

#训练SGD
print('=' * 80)
print("Elastic-Net penalty")
results.append(benchmark(SGDClassifier(alpha=.0001, max_iter=50,
                                       penalty="elasticnet")))

#无阈值训练
print('=' * 80)
print("NearestCentroid (aka Rocchio classifier)")
results.append(benchmark(NearestCentroid()))

#训练稀疏朴素贝叶斯分类器
print('=' * 80)
print("Naive Bayes")
results.append(benchmark(MultinomialNB(alpha=.01)))
results.append(benchmark(BernoulliNB(alpha=.01)))
results.append(benchmark(ComplementNB(alpha=.1)))

print('=' * 80)
print("LinearSVC with L1-based feature selection")
#C越小,正则化越强
```

```
#正则化越多,稀疏性越强
results.append(benchmark(Pipeline([
  ('feature_selection', SelectFromModel(LinearSVC(penalty="l1", dual=False,
                                                  tol=1e-3))),
  ('classification', LinearSVC(penalty="l2"))])))
```

（3）开始绘图，条形图分别表示每个分类器的准确度、训练时间（归一化）和测试时间（归一化）。代码如下：

```
indices = np.arange(len(results))

results = [[x[i] for x in results] for i in range(4)]

clf_names, score, training_time, test_time = results
training_time = np.array(training_time) / np.max(training_time)
test_time = np.array(test_time) / np.max(test_time)

plt.figure(figsize=(12, 8))
plt.title("Score")
plt.barh(indices, score, .2, label="score", color='navy')
plt.barh(indices + .3, training_time, .2, label="training time",
         color='c')
plt.barh(indices + .6, test_time, .2, label="test time", color='darkorange')
plt.yticks(())
plt.legend(loc='best')
plt.subplots_adjust(left=.25)
plt.subplots_adjust(top=.95)
plt.subplots_adjust(bottom=.05)

for i, c in zip(indices, clf_names):
    plt.text(-.3, i, c)

plt.show()
```

linear10.py 执行后的效果如图 3-11 所示。

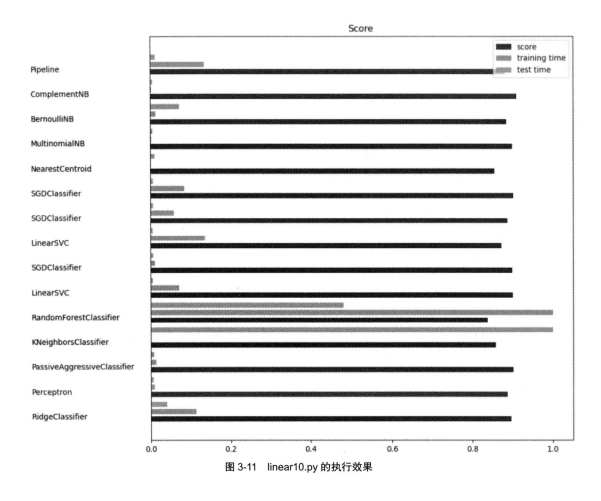

图 3-11 linear10.py 的执行效果

第 4 章

无监督学习

利用无标签的数据学习数据的分布或数据与数据之间的关系被称作无监督学习,有监督学习和无监督学习的最大区别在于数据是否有标签。在本章的内容中,将详细介绍使用 Scikit-Learn 实现无监督学习的知识,为读者步入本书后面知识的学习打下基础。

4.1 高斯混合模型

混合模型是一个可以用来表示在总体分布（distribution）中含有 K 个子分布的概率模型，换句话说，混合模型表示观测数据在总体中的概率分布，它是一个由 K 个子分布组成的混合分布。混合模型不要求观测数据提供关于子分布的信息来计算观测数据在总体分布中的概率。

扫码观看本节视频讲解

高斯混合模型可以看作由 K 个单高斯模型组合而成的模型，这 K 个子模型是混合模型的隐变量（hidden variable）。一般来说，一个混合模型可以使用任何概率分布，这里使用高斯混合模型是因为高斯分布具备良好的数学性质及计算性能。

比如我们现在有一组狗的样本数据，不同种类的狗，其体型、颜色、长相各不相同，但都属于狗这个种类，此时利用单高斯模型不能很好地来描述这个分布，因为样本数据分布并不是一个单一的椭圆，所以用混合高斯分布可以更好地描述这个问题，如图 4-1 所示。

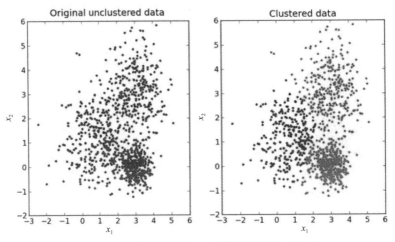

图 4-1　图中的每个点都由 K 个子模型中的某一个生成

4.1.1 高斯混合

在 Scikit-Learn 应用中，sklearn.mixture 是一个应用高斯混合模型进行无监督学习的包，支持 diagonal、spherical、tied、full 四种协方差矩阵。
- diagonal：指每个分量有各自不同的对角协方差矩阵；
- spherical：指每个分量有各自不同的简单协方差矩阵；
- tied：指所有分量有相同的标准协方差矩阵；
- full：指每个分量有各自不同的标准协方差矩阵。

sklearn.mixture 对数据进行抽样，并且根据数据估计模型。同时无监督学习的包也提供了相关支持，来帮助用户决定合适的分量数（分量个数）。在高斯混合模型中，将每一个高斯分布称为一个分量，即 component。

在 Scikit-Learn 中，GaussianMixture 对象实现了用来拟合高斯混合模型的期望最大化（EM）算法。GaussianMixture 还可以为多变量模型绘制置信区间，同时计算贝叶斯信息准则（bayesian information criterion, BIC）来评估数据中聚类的数量。GaussianMixture.fit 提供了从训练数据中学习高斯混合模型的方法。

如果提供了测试数据，通过使用方法 GaussianMixture.predict() 可以为每个样本分配最有可能对应的高斯分布。在 GaussianMixture 方法中自带了不同的选项来约束不同的估计协方差：diagonal、spherical、tied 或 full 协方差。

在现实应用中，经常使用贝叶斯信息准则来选择高斯混合的分量数。理论上，它仅当在近似状态下可以恢复正确的分量数（如果有大量数据可用，并且假设这些数据实际上是一个混合高斯模型独立分布生成的）。注意，使用变分贝叶斯高斯混合可以避免高斯混合模型中分量数的选择。

请看下面的实例文件 gaussian01.py，这是一个用典型的高斯混合模型进行选择的例子。本实例使用高斯混合模型对 BIC 执行模型选择。模型选择涉及协方差类型和模型中的组件数量。在这种情况下，AIC（Akaike information Criterion，赤池信息量准则，是衡量统计模型拟合优良性的一种标准）也会提供正确的结果（为了节省时间本实例未显示），但如果是识别正确的模型，则 BIC 更适合。在这种情况下，可选择具有两个分量和完全协方差的模型（对应于真正的生成模型）。

```
#每个组件的样本数
n_samples = 500

#生成随机样本,两个分量
np.random.seed(0)
C = np.array([[0., -0.1], [1.7, .4]])
X = np.r_[np.dot(np.random.randn(n_samples, 2), C),
          .7 * np.random.randn(n_samples, 2) + np.array([-6, 3])]

lowest_bic = np.infty
bic = []
n_components_range = range(1, 7)
cv_types = ['spherical', 'tied', 'diag', 'full']
for cv_type in cv_types:
    for n_components in n_components_range:
        #用EM拟合高斯混合
        gmm = mixture.GaussianMixture(n_components=n_components,
                                      covariance_type=cv_type)
        gmm.fit(X)
        bic.append(gmm.bic(X))
        if bic[-1] < lowest_bic:
            lowest_bic = bic[-1]
            best_gmm = gmm

bic = np.array(bic)
color_iter = itertools.cycle(['navy', 'turquoise', 'cornflowerblue',
```

```python
                                'darkorange'])
clf = best_gmm
bars = []

#绘制BIC分数
plt.figure(figsize=(8, 6))
spl = plt.subplot(2, 1, 1)
for i, (cv_type, color) in enumerate(zip(cv_types, color_iter)):
    xpos = np.array(n_components_range) + .2 * (i - 2)
    bars.append(plt.bar(xpos, bic[i * len(n_components_range):
                        (i + 1) * len(n_components_range)],
                        width=.2, color=color))
plt.xticks(n_components_range)
plt.ylim([bic.min() * 1.01 - .01 * bic.max(), bic.max()])
plt.title('BIC score per model')
xpos = np.mod(bic.argmin(), len(n_components_range)) + .65 +\
    .2 * np.floor(bic.argmin() / len(n_components_range))
plt.text(xpos, bic.min() * 0.97 + .03 * bic.max(), '*', fontsize=14)
spl.set_xlabel('Number of components')
spl.legend([b[0] for b in bars], cv_types)

#绘制得分
splot = plt.subplot(2, 1, 2)
Y_ = clf.predict(X)
for i, (mean, cov, color) in enumerate(zip(clf.means_, clf.covariances_,
                                            color_iter)):
    v, w = linalg.eigh(cov)
    if not np.any(Y_ == i):
        continue
    plt.scatter(X[Y_ == i, 0], X[Y_ == i, 1], .8, color=color)

    #绘制一个椭圆以显示高斯分量
    angle = np.arctan2(w[0][1], w[0][0])
    angle = 180. * angle / np.pi   #转换为度
    v = 2. * np.sqrt(2.) * np.sqrt(v)
    ell = mpl.patches.Ellipse(mean, v[0], v[1], 180. + angle, color=color)
    ell.set_clip_box(splot.bbox)
    ell.set_alpha(.5)
    splot.add_artist(ell)

plt.xticks(())
plt.yticks(())
plt.title(f'Selected GMM: {best_gmm.covariance_type} model, '
          f'{best_gmm.n_components} components')
plt.subplots_adjust(hspace=.35, bottom=.02)
plt.show()
```

gaussian01.py 执行后的效果如图 4-2 所示。

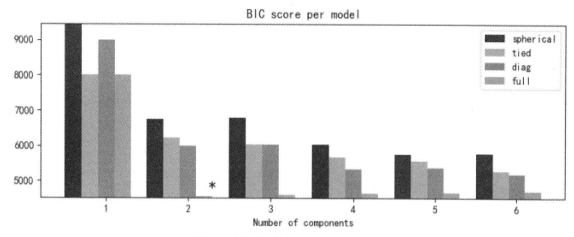

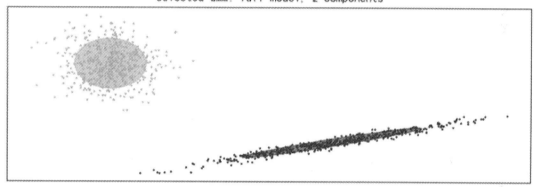

图 4-2　gaussian01.py 的执行效果

4.1.2　变分贝叶斯高斯混合

在 Scikit-Learn 应用中，变分贝叶斯高斯混合（BayesianGaussianMixture）对象实现了具有变分的高斯混合模型的变体推理算法，这个 API 的功能和混合高斯（GaussianMixture）相似。

变分推断是期望最大化（EM）的扩展，它最大化模型证据（包括先验）的下界，而不是数据似然函数。 变分方法的原理与期望最大化相同（二者都是迭代算法，在寻找由混合产生的每个点的概率和根据所分配的点拟合之间两步交替），但是变分方法通过整合先验分布信息来增加正则化限制。 这避免了期望最大化解决方案中常出现的奇异性，但是也对模型带来了微小的偏差。 变分方法计算过程通常明显较慢，但不会慢到无法使用。

由于它的贝叶斯特性，变分算法比期望最大化（EM）需要更多的超参数（先验分布中的参数），其中最重要的就是浓度参数 weight_concentration_prior。指定一个低浓度先验，将会使模型大部分的权重放在少数分量上，其余分量的权重则趋近 0。而高浓度先验将使混合模型中的大部分分量都有一定的权重。

BayesianGaussianMixture 类的参数实现提出了两种权重分布先验：一种是利用狄利克雷分布（Dirichlet distribution）的有限混合模型，另一种是利用狄利克雷过程（Dirichlet process）的无限混合模型。在实际应用上，狄利克雷过程推理算法是近似的，并且使用具有固定最大分量数的截尾分布（称之为 Stick-breaking 表示法）。使用的分量数实际上几乎是取决于数据。

请看下面的实例文件 gaussian02.py，功能是使用 GaussianMixture 和 BayesianGaussianMixture 绘制置信椭圆体。使用期望最大化（GaussianMixture 类）和变分推理（BayesianGaussianMixture，先验狄利克雷过程的类模型）绘制获得的两个高斯混合的置信椭圆体。

```
color_iter = itertools.cycle(['navy', 'c', 'cornflowerblue', 'gold',
                              'darkorange'])

def plot_results(X, Y_, means, covariances, index, title):
    splot = plt.subplot(2, 1, 1 + index)
    for i, (mean, covar, color) in enumerate(zip(
            means, covariances, color_iter)):
        v, w = linalg.eigh(covar)
        v = 2. * np.sqrt(2.) * np.sqrt(v)
        u = w[0] / linalg.norm(w[0])
        #由于DP不会使用它可以访问的所有组件,除非必须使用,否则不应该绘制冗余组件
        if not np.any(Y_ == i):
            continue
        plt.scatter(X[Y_ == i, 0], X[Y_ == i, 1], .8, color=color)

        #绘制一个椭圆以显示高斯分量
        angle = np.arctan(u[1] / u[0])
        angle = 180. * angle / np.pi   #转换为度
        ell = mpl.patches.Ellipse(mean, v[0], v[1], 180. + angle, color=color)
        ell.set_clip_box(splot.bbox)
        ell.set_alpha(0.5)
        splot.add_artist(ell)

    plt.xlim(-9., 5.)
    plt.ylim(-3., 6.)
    plt.xticks(())
    plt.yticks(())
    plt.title(title)

#每个组件的样本数
n_samples = 500

#生成随机样本,两个分量
np.random.seed(0)
C = np.array([[0., -0.1], [1.7, .4]])
X = np.r_[np.dot(np.random.randn(n_samples, 2), C),
          .7 * np.random.randn(n_samples, 2) + np.array([-6, 3])]
```

```
#用五个分量拟合高斯混合电磁波
gmm = mixture.GaussianMixture(n_components=5, covariance_type='full').fit(X)
plot_results(X, gmm.predict(X), gmm.means_, gmm.covariances_, 0,
             'Gaussian Mixture')

#用五个分量拟合Dirichlet过程高斯混合
dpgmm = mixture.BayesianGaussianMixture(n_components=5,
                                        covariance_type='full').fit(X)
plot_results(X, dpgmm.predict(X), dpgmm.means_, dpgmm.covariances_, 1,
             'Bayesian Gaussian Mixture with a Dirichlet process prior')

plt.show()
```

在本实例中，两种模型都可以访问五个用于拟合数据的组件。请注意，期望最大化模型必须使用所有五个组件，而变分推理模型将有效地仅使用良好拟合所需的数量。在这里我们可以看到，期望最大化模型任意拆分了一些组件，因为它试图拟合更多组件，而狄利克雷过程模型会自动适应它的状态数。

4.2 流形学习

流形学习是一种非线性降维方法，其算法基于的思想是：许多数据集的维度过高是由人为导致的。高维数据集非常难以可视化，虽然可以绘制二维或三维的数据来显示数据的固有结构，但是与之等效的高维图却不太直观。为了帮助数据集结构的可视化，必须以某种方式降低维度。

扫码观看本节视频讲解

通过对数据的随机投影来实现降维是最简单的方法，虽然这样做能实现数据结构一定程度上的可视化，但随机选择投影仍有许多待改进之处。在随机投影中，很可能会丢失数据中更有趣的结构。为了解决这一问题，一些监督和无监督的线性降维框架被设计出来，如主成分分析（PCA）、独立成分分析、线性判别分析等。这些算法定义了明确的规定来选择数据的"有趣的"线性投影。它们虽然强大，但是会经常错失数据中重要的非线性结构。

可以将流形学习认为是一种将线性框架（如 PCA）推广为对数据中非线性结构敏感的尝试，虽然存在监督变量，但是典型的流形学习问题是无监督的：它从数据本身学习数据的高维结构，而不使用预定的分类。

请看下面的实例文件 gaussian03.py，功能是使用各种流形学习方法对 S 曲线数据集进行降维。请注意，多维度分析的目的是找到数据的低维表示（此处为二维），其中曲线振幅贴合原始高维空间中的距离，与其他流形学习算法不同，它不寻求低维空间中数据的各向同性表示。

```
Axes3D

n_points = 1000
X, color = datasets.make_s_curve(n_points, random_state=0)
n_neighbors = 10
n_components = 2

#创建图形
```

```python
fig = plt.figure(figsize=(15, 8))
fig.suptitle("Manifold Learning with %i points, %i neighbors"
             % (1000, n_neighbors), fontsize=14)

#添加三维散点图
ax = fig.add_subplot(251, projection='3d')
ax.scatter(X[:, 0], X[:, 1], X[:, 2], c=color, cmap=plt.cm.Spectral)
ax.view_init(4, -72)

#设置多种方法
LLE = partial(manifold.LocallyLinearEmbedding,
              n_neighbors, n_components, eigen_solver='auto')

methods = OrderedDict()
methods['LLE'] = LLE(method='standard')
methods['LTSA'] = LLE(method='ltsa')
methods['Hessian LLE'] = LLE(method='hessian')
methods['Modified LLE'] = LLE(method='modified')
methods['Isomap'] = manifold.Isomap(n_neighbors, n_components)
methods['MDS'] = manifold.MDS(n_components, max_iter=100, n_init=1)
methods['SE'] = manifold.SpectralEmbedding(n_components=n_components,
                                           n_neighbors=n_neighbors)
methods['t-SNE'] = manifold.TSNE(n_components=n_components, init='pca',
                                 random_state=0)

#绘制结果
for i, (label, method) in enumerate(methods.items()):
    t0 = time()
    Y = method.fit_transform(X)
    t1 = time()
    print("%s: %.2g sec" % (label, t1 - t0))
    ax = fig.add_subplot(2, 5, 2 + i + (i > 3))
    ax.scatter(Y[:, 0], Y[:, 1], c=color, cmap=plt.cm.Spectral)
    ax.set_title("%s (%.2g sec)" % (label, t1 - t0))
    ax.xaxis.set_major_formatter(NullFormatter())
    ax.yaxis.set_major_formatter(NullFormatter())
    ax.axis('tight')

plt.show()
```

gaussian03.py 的执行效果如图 4-3 所示。

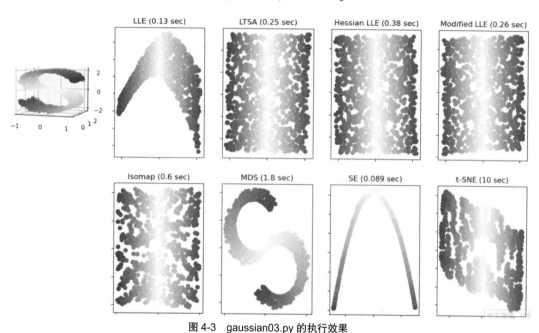

图 4-3 gaussian03.py 的执行效果

4.3 聚类

在 Scikit-Learn 应用中,可以使用模块 sklearn.cluster 来实现未标记的数据聚类(clusters)。每个聚类算法(clustering algorithm)有如下两个变体。

- 类(class):实现方法 fit() 来学习训练数据(train data)的聚类(clusters)。
- 函数(function):用于提供训练数据,返回与不同 clusters 对应的整数标签 array(数组)。

扫码观看本节视频讲解

需要注意的一点是,在模块 sklearn.cluster 中的算法可以采用不同种类的 matrix(矩阵)作为输入。所有这些都接受 shape [n_samples, n_features] 的标准数据矩阵,这些可以从 sklearn.feature_extraction 模块的 classes(类)中获得。对于 AffinityPropagation(AP 聚类算法)来说,SpectralClustering 和 DBSCAN 也可以输入 shape [n_samples, n_features] 的相似矩阵,这些可以从 sklearn.metrics.pairwise 模块中的函数获得。

4.3.1 KMeans 算法

在 Scikit-Learn 应用中,KMeans(K-均值)聚类算法通过试图分离 n 个相等方差组(n groups of equal variance)的样本来聚集数据,最小化(minimizing)称为惯性(inertia)或者簇内和平方(within-cluster sum-of-squares）的标准（criterion）。该算法需要指定簇的数量（number of clusters）。它可以很好地扩展（scales）到大量样本（large number of samples），并已经被广泛应用于许多不同的领域。

KMeans 算法将一组 N 样本 X 划分成 K 不相交的簇（clusters），每个 cluster 都用簇中样本的均值 μ_j 来描述。这个均值（mean）通常被称为簇的质心（centroids）。

注意，一般不是从 X 中挑选出的点，虽然它们是处在同一个空间（space）。KMeans 算法旨在选择最小化惯性（inertia）或簇内和的平方和（within-cluster sum of squared）的标准的 centroids：

$$\sum_{i=0}^{n} \min_{\mu_j \in C} \left(\left\| x_j - \mu_i \right\|^2 \right)$$

请看下面的实例文件 gaussian04.py，功能是说明 KMeans 是否直观执行、何时不执行的情况。

```python
import numpy as np
import matplotlib.pyplot as plt

from sklearn.cluster import KMeans
from sklearn.datasets import make_blobs

plt.figure(figsize=(12, 12))

n_samples = 1500
random_state = 170
X, y = make_blobs(n_samples=n_samples, random_state=random_state)

#簇数不正确
y_pred = KMeans(n_clusters=2, random_state=random_state).fit_predict(X)

plt.subplot(221)
plt.scatter(X[:, 0], X[:, 1], c=y_pred)
plt.title("Incorrect Number of Blobs")

#各向异性分布数据
transformation = [[0.60834549, -0.63667341], [-0.40887718, 0.85253229]]
X_aniso = np.dot(X, transformation)
y_pred = KMeans(n_clusters=3, random_state=random_state).fit_predict(X_aniso)

plt.subplot(222)
plt.scatter(X_aniso[:, 0], X_aniso[:, 1], c=y_pred)
plt.title("Anisotropicly Distributed Blobs")

#差异
X_varied, y_varied = make_blobs(n_samples=n_samples,
                                cluster_std=[1.0, 2.5, 0.5],
                                random_state=random_state)
y_pred = KMeans(n_clusters=3, random_state=random_state).fit_predict(X_varied)

plt.subplot(223)
plt.scatter(X_varied[:, 0], X_varied[:, 1], c=y_pred)
```

```
plt.title("Unequal Variance")

#大小不均的斑点
X_filtered = np.vstack((X[y == 0][:500], X[y == 1][:100], X[y == 2][:10]))
y_pred = KMeans(n_clusters=3,
                random_state=random_state).fit_predict(X_filtered)

plt.subplot(224)
plt.scatter(X_filtered[:, 0], X_filtered[:, 1], c=y_pred)
plt.title("Unevenly Sized Blobs")

plt.show()
```

gaussian04.py 执行效果如图 4-4 所示,在前三个图中,输入数据不符合 KMeans 做出的一些隐含假设,结果产生了不需要的集群。在最后一个图中,尽管点大小不均匀,但 KMeans 返回了直观的集群。

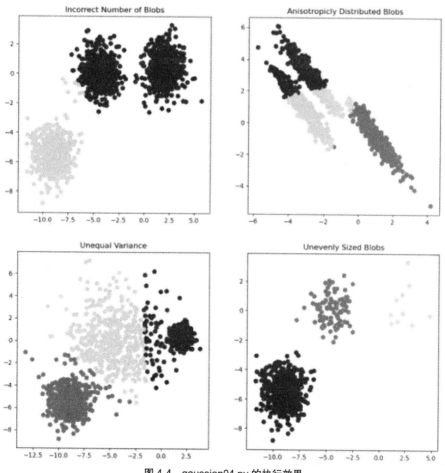

图 4-4　gaussian04.py 的执行效果

4.3.2 MiniBatchKMeans 算法

MiniBatchKMeans 算法是 KMeans 算法的一个变体，它使用小批量（Mini-batches）来减少计算时间，同时仍然尝试优化相同的目标函数（objective function）。小批量是输入数据的子集，在每次训练迭代（training iteration）中随机抽样（randomly sampled）。这些小批量大大减少了融合到本地解决方案所需的计算量。与其他降低 KMeans 收敛时间的算法相反，MiniBatchKMeans 产生的结果通常只比标准算法略差。

MiniBatchKMeans 算法在两个主要步骤之间进行迭代，类似于 vanilla KMeans。在第一步，b 样本是从数据集中随机抽取的，形成一个小批量。然后将它们分配到最近的质心（centroid）。在第二步，质心被更新。与 KMeans 相反，这是在每个样本的基础上完成的。对于小批量中的每个样本，通过取样本的流平均值（streaming average）和分配给该质心的所有先前样本来更新分配的质心。这具有随时间降低质心的变化率（rate）的效果。执行这些步骤直到收敛或达到预定次数的迭代。MiniBatchKMeans 收敛速度比 KMeans 快，但是结果的质量会降低。在实践应用中，质量差异可能会相当小。

请看下面的实例文件 gaussian05.py，功能是比较 KMeans 算法与 MiniBatchKMeans 算法。本实例将对一组数据进行聚类，首先使用 KMeans，然后使用 MiniBatchKMeans，绘制结果，并绘制两种算法之间的不同点。

```
#生成示例数据
np.random.seed(0)

batch_size = 45
centers = [[1, 1], [-1, -1], [1, -1]]
n_clusters = len(centers)
X, labels_true = make_blobs(n_samples=3000, centers=centers, cluster_std=0.7)


k_means = KMeans(init='k-means++', n_clusters=3, n_init=10)
t0 = time.time()
k_means.fit(X)
t_batch = time.time() - t0

# #############################################################################
mbk = MiniBatchKMeans(init='k-means++', n_clusters=3, batch_size=batch_size,
                      n_init=10, max_no_improvement=10, verbose=0)
t0 = time.time()
mbk.fit(X)
t_mini_batch = time.time() - t0


# #############################################################################
#绘制

fig = plt.figure(figsize=(8, 3))
fig.subplots_adjust(left=0.02, right=0.98, bottom=0.05, top=0.9)
colors = ['#4EACC5', '#FF9C34', '#4E9A06']
```

```python
#我们希望MiniBatchKMeans和KMeans算法中的同一簇具有相同的颜色,让我们把每个最近的中心配对
k_means_cluster_centers = k_means.cluster_centers_
order = pairwise_distances_argmin(k_means.cluster_centers_,
                                  mbk.cluster_centers_)
mbk_means_cluster_centers = mbk.cluster_centers_[order]

k_means_labels = pairwise_distances_argmin(X, k_means_cluster_centers)
mbk_means_labels = pairwise_distances_argmin(X, mbk_means_cluster_centers)

#K-Means
ax = fig.add_subplot(1, 3, 1)
for k, col in zip(range(n_clusters), colors):
    my_members = k_means_labels == k
    cluster_center = k_means_cluster_centers[k]
    ax.plot(X[my_members, 0], X[my_members, 1], 'w',
            markerfacecolor=col, marker='.')
    ax.plot(cluster_center[0], cluster_center[1], 'o', markerfacecolor=col,
            markeredgecolor='k', markersize=6)
ax.set_title('KMeans')
ax.set_xticks(())
ax.set_yticks(())
plt.text(-3.5, 1.8,  'train time: %.2fs\ninertia: %f' % (
    t_batch, k_means.inertia_))

#MiniBatchKMeans
ax = fig.add_subplot(1, 3, 2)
for k, col in zip(range(n_clusters), colors):
    my_members = mbk_means_labels == k
    cluster_center = mbk_means_cluster_centers[k]
    ax.plot(X[my_members, 0], X[my_members, 1], 'w',
            markerfacecolor=col, marker='.')
    ax.plot(cluster_center[0], cluster_center[1], 'o', markerfacecolor=col,
            markeredgecolor='k', markersize=6)
ax.set_title('MiniBatchKMeans')
ax.set_xticks(())
ax.set_yticks(())
plt.text(-3.5, 1.8, 'train time: %.2fs\ninertia: %f' %
         (t_mini_batch, mbk.inertia_))

#将不同的数组初始化为False
different = (mbk_means_labels == 4)
ax = fig.add_subplot(1, 3, 3)

for k in range(n_clusters):
    different += ((k_means_labels == k) != (mbk_means_labels == k))

identic = np.logical_not(different)
ax.plot(X[identic, 0], X[identic, 1], 'w',
```

```
            markerfacecolor='#bbbbbb', marker='.')
ax.plot(X[different, 0], X[different, 1], 'w',
            markerfacecolor='m', marker='.')
ax.set_title('Difference')
ax.set_xticks(())
ax.set_yticks(())

plt.show()
```

gaussian05.py 的执行效果如图 4-5 所示。

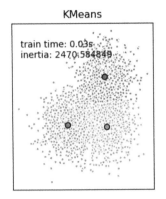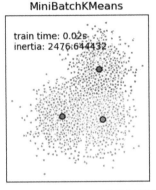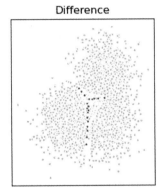

图 4-5　gaussian05.py 的执行效果

4.4　双聚类

在 Scikit-Learn 应用中，可以使用 sklearn.cluster.bicluster 模块实现双聚类（Biclustering）。Biclustering 算法对数据矩阵的行列同时进行聚类，将同时对行和列进行的聚类称为双向聚类。每一次聚类都会通过原始数据矩阵的一些属性确定一个子矩阵。例如，一个矩阵（10,10），一个 Biclustering 聚类有三行二列，就可以用一个子矩阵（3,2）：

扫码观看本节视频讲解

```
>>> import numpy as np
>>> data = np.arange(100).reshape(10, 10)
>>> rows = np.array([0, 2, 3])[:, np.newaxis]
>>> columns = np.array([1, 2])
>>> data[rows, columns]
array([[ 1,  2],
       [21, 22],
       [31, 32]])
```

为了实现可视化效果，设置一个 Biclustering 双聚类，数据矩阵的行列可以重新分配，使 bicluster 是连续的。

4.4.1 谱聚类算法

谱聚类算法（Spectral Clustering）找到的 bicluster 值比相应的其他行和列更高，因为每一行和每一列都只属于一个 bicluster，所以重新分配行和列，使分区连续显示对角线上的 high 值。

找到最优归一化剪切的近似解，可以通过图形的 Laplacian 的广义特征值分解。通常这意味着可以直接使用 Laplacian 矩阵。如果原始数据矩阵 A 有形状 $m*n$，则对应的 bipartite 图的 Laplacian 矩阵具有形状 $(m+n)*(m+n)$。但是，在这种情况下建议直接使用 A，因为它更小，更有效。

输入矩阵 A 被预处理为：

$$A_n = R^{-1/2} A C^{-1/2}$$

其中，R 是 i 的对角线矩阵，和 $\sum_j A_{ij}$ 相同，C 是 j 的对角吸纳矩阵，等同于 $\sum_j A_{ij} A\_\{ij\}$。然后进行奇异值分解，$A\_n = A_n = U\sum V^T$ 提供了 A 行列的分区。左边的奇异值向量给予行分区，右边的奇异值向量给予列分区。

$e = \lceil \log_2 k \rceil$ 奇异值向量从第二个开始，提供所需的分区信息。这些用于形成矩阵 Z：

$$Z = \begin{bmatrix} R^{-1/2}U \\ C^{-1/2}V \end{bmatrix}$$

U 的列是 $u_2, \cdots, u_{e+1}$，和 V 相似。然后 Z 的 rows 通过使用 KMeans 进行聚类。n_rows 标签提供行分区，剩下的 n_columns 标签提供列分区。

请看下面的实例文件 gaussian06.py，演示了在 20 个新闻组数据集上使用 Spectral Clustering 算法的过程。在数据集中，将 comp.os.ms-windows.misc 类别排除在外，因为它包含许多只包含数据的帖子。各个 TF-IDF 矢量化帖子形成一个词频矩阵，然后使用 Dhillon 的 Spectral Clustering 算法对其进行双聚类处理。生成的文档词双簇表示在这些子集文档中被经常使用的子集词。

```
def number_normalizer(tokens):
    """ 将所有数字标记映射到占位符.
    对于许多应用程序,以数字开头的令牌并不是直接有用的,但是这样一个令牌的存在可能是相关的.
    通过应用这种形式的降维,某些方法的性能可能会更好
    """
    return ("#NUMBER" if token[0].isdigit() else token for token in tokens)


class NumberNormalizingVectorizer(TfidfVectorizer):
    def build_tokenizer(self):
        tokenize = super().build_tokenizer()
        return lambda doc: list(number_normalizer(tokenize(doc)))


# exclude 'comp.os.ms-windows.misc'
categories = ['alt.atheism', 'comp.graphics',
              'comp.sys.ibm.pc.hardware', 'comp.sys.mac.hardware',
              'comp.windows.x', 'misc.forsale', 'rec.autos',
              'rec.motorcycles', 'rec.sport.baseball',
```

```python
                    'rec.sport.hockey', 'sci.crypt', 'sci.electronics',
                    'sci.med', 'sci.space', 'soc.religion.christian',
                    'talk.politics.guns', 'talk.politics.mideast',
                    'talk.politics.misc', 'talk.religion.misc']
newsgroups = fetch_20newsgroups(categories=categories)
y_true = newsgroups.target

vectorizer = NumberNormalizingVectorizer(stop_words='english', min_df=5)
cocluster = SpectralCoclustering(n_clusters=len(categories),
                                 svd_method='arpack', random_state=0)
kmeans = MiniBatchKMeans(n_clusters=len(categories), batch_size=20000,
                        random_state=0)

print("Vectorizing...")
X = vectorizer.fit_transform(newsgroups.data)

print("Coclustering...")
start_time = time()
cocluster.fit(X)
y_cocluster = cocluster.row_labels_
print("Done in {:.2f}s. V-measure: {:.4f}".format(
    time() - start_time,
    v_measure_score(y_cocluster, y_true)))

print("MiniBatchKMeans...")
start_time = time()
y_kmeans = kmeans.fit_predict(X)
print("Done in {:.2f}s. V-measure: {:.4f}".format(
    time() - start_time,
    v_measure_score(y_kmeans, y_true)))

feature_names = vectorizer.get_feature_names()
document_names = list(newsgroups.target_names[i] for i in newsgroups.target)


def bicluster_ncut(i):
    rows, cols = cocluster.get_indices(i)
    if not (np.any(rows) and np.any(cols)):
        import sys
        return sys.float_info.max
    row_complement = np.nonzero(np.logical_not(cocluster.rows_[i]))[0]
    col_complement = np.nonzero(np.logical_not(cocluster.columns_[i]))[0]
    #注意:以下内容与X[rows[:, np.newaxis], cols].sum()相同,但在scipy中要快得多,小于等于0.16
    weight = X[rows][:, cols].sum()
    cut = (X[row_complement][:, cols].sum() +
           X[rows][:, col_complement].sum())
    return cut / weight
```

```python
def most_common(d):
    """defaultdict(int)中具有最高值的项,类似于Counter.most_common
    """
    return sorted(d.items(), key=operator.itemgetter(1), reverse=True)


bicluster_ncuts = list(bicluster_ncut(i)
                       for i in range(len(newsgroups.target_names)))
best_idx = np.argsort(bicluster_ncuts)[:5]

print()
print("Best biclusters:")
print("----------------")
for idx, cluster in enumerate(best_idx):
    n_rows, n_cols = cocluster.get_shape(cluster)
    cluster_docs, cluster_words = cocluster.get_indices(cluster)
    if not len(cluster_docs) or not len(cluster_words):
        continue

    #类别
    counter = defaultdict(int)
    for i in cluster_docs:
        counter[document_names[i]] += 1
    cat_string = ", ".join("{:.0f}% {}".format(float(c) / n_rows * 100, name)
                           for name, c in most_common(counter)[:3])

    #单词
    out_of_cluster_docs = cocluster.row_labels_ != cluster
    out_of_cluster_docs = np.where(out_of_cluster_docs)[0]
    word_col = X[:, cluster_words]
    word_scores = np.array(word_col[cluster_docs, :].sum(axis=0) -
                           word_col[out_of_cluster_docs, :].sum(axis=0))
    word_scores = word_scores.ravel()
    important_words = list(feature_names[cluster_words[i]]
                           for i in word_scores.argsort()[:-11:-1])

    print("bicluster {} : {} documents, {} words".format(
        idx, n_rows, n_cols))
    print("categories   : {}".format(cat_string))
    print("words        : {}\n".format(', '.join(important_words)))
```

为了进行对比,在上述代码中还使用 MiniBatchKMeans 算法对文档进行了聚类处理。从双聚类导出的文档聚类比 MiniBatchKMeans 发现的聚类实现了更好的 V-measure。执行后会输出:

```
Vectorizing...
Coclustering...
Done in 2.18s. V-measure: 0.4431
MiniBatchKMeans...
Done in 7.83s. V-measure: 0.3344
```

```
Best biclusters:
----------------
bicluster 0  : 1961 documents, 4388 words
categories   : 23% talk.politics.guns, 18% talk.politics.misc, 17% sci.med
words        : gun, geb, guns, banks, gordon, clinton, pitt, cdt, surrender, veal

bicluster 1  : 1269 documents, 3558 words
categories   : 27% soc.religion.christian, 25% talk.politics.mideast, 24% alt.atheism
words        : god, jesus, christians, sin, objective, kent, belief, christ, faith, moral

bicluster 2: 2201 documents, 2747 words
categories   : 18% comp.sys.mac.hardware, 17% comp.sys.ibm.pc.hardware, 16% comp.graphics
words        : voltage, board, dsp, packages, receiver, stereo, shipping, package, compression, image

bicluster 3  : 1773 documents, 2620 words
categories   : 27% rec.motorcycles, 23% rec.autos, 13% misc.forsale
words        : bike, car, dod, engine, motorcycle, ride, honda, bikes, helmet, bmw

bicluster 4  : 201 documents, 1175 words
categories   : 81% talk.politics.mideast, 10% alt.atheism, 7% soc.religion.christian
words        : turkish, armenia, armenian, armenians, turks, petch, sera, zuma, argic, gvg47
```

4.4.2 光谱联合聚类算法

光谱联合聚类算法（Spectral Co-Clustering）找到的值高于相应的其他行和列中的值，每行和每列只属于一个双聚类，因此重新排列行和列中的这些高值，使这些分区沿着对角线连续显示。

在 Scikit-Learn 应用中，实现光谱联合聚类算法的函数是 sklearn.cluster.bicluster.SpectralCoclustering，主要参数的具体说明如下。

- n_clusters：聚类中心的数目，默认是3。
- svd_method：计算singular vectors的算法——randomized（默认）或arpack。
- n_svd_vecs：计算singular vectors值时使用的向量数目。
- n_jobs：计算时采用的线程或进程数量。

在 Scikit-Learn 应用中，实现光谱联合聚类算法的主要属性如下。

- rows_：二维数组，表示聚类的结果。其中的值都是True或False。如果rows_[i,r]为True，表示聚类i包含行r。
- columns_：二维数组，表示聚类的结果。
- row_labels_：每行的聚类标签列表。
- column_labels_：每列的聚类标签列表。

请看下面的实例文件 gaussian07.py，功能是使用光谱联合聚类算法生成数据集并对其进行双聚类处理。

```
data, rows, columns = make_biclusters(
```

```
    shape=(300, 300), n_clusters=5, noise=5,
    shuffle=False, random_state=0)

plt.matshow(data, cmap=plt.cm.Blues)
plt.title("Original dataset")

#洗牌群集
rng = np.random.RandomState(0)
row_idx = rng.permutation(data.shape[0])
col_idx = rng.permutation(data.shape[1])
data = data[row_idx][:, col_idx]

plt.matshow(data, cmap=plt.cm.Blues)
plt.title("Shuffled dataset")

model = SpectralCoclustering(n_clusters=5, random_state=0)
model.fit(data)
score = consensus_score(model.biclusters_,
                        (rows[:, row_idx], columns[:, col_idx]))

print("consensus score: {:.3f}".format(score))

fit_data = data[np.argsort(model.row_labels_)]
fit_data = fit_data[:, np.argsort(model.column_labels_)]

plt.matshow(fit_data, cmap=plt.cm.Blues)
plt.title("After biclustering; rearranged to show biclusters")

plt.show()
```

在上述代码中，使用函数 make_biclusters() 生成数据集，该函数创建一个小值矩阵并植入大值的双簇。然后将行和列打乱并传递给光谱联合聚类算法。重新排列混淆矩阵，以连续显示 biclusters 算法找到双聚类的准确程度。执行后的效果如图 4-6 所示，3 个可视化图依次表示原始数据、打乱后的数据和聚类后的效果图。

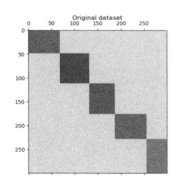
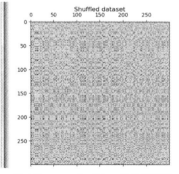
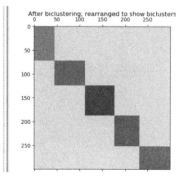

图 4-6　gaussian07.py 的执行效果

第 5 章

模型选择和评估

在本书前面介绍了监督学习算法和无监督学习算法的知识,当我们使用这些算法制作模型时,应该如何挑选使用某一种具体算法呢?我们需要先比较模型的性能,然后再做出选择,这就需要用到模型评估与选择的知识。在本章的内容中,将详细介绍基于 Scikit-Learn 实现模型选择和评估的知识,为读者步入本书后面知识的学习打下基础。

5.1 交叉验证：评估估算器的表现

在学习一个数据集中的信息后，在相同数据集上进行测试是一种错误的做法。这样一个仅给出测试用例标签的模型将会获得极高的分数，但对于尚未出现过的数据则无法预测出任何有用的信息，我们将这种情况称为过拟合（overfitting）。为了避免这种情况发生，在进行（监督）机器学习实验时，通常取出部分可利用数据作为测试数据集：X_test 和 y_test。

扫码观看本节视频讲解

在 Scikit-Learn 应用中，使用 train_test_split 库中的辅助函数可以很快地将测试数据集划分为任何训练集 (training sets) 和测试集（test sets）。例如，在下面的代码中加载了 iris 数据集，并在此数据集上训练出线性支持向量机。

```
>>> import numpy as np
>>> from sklearn.model_selection import train_test_split
>>> from sklearn import datasets
>>> from sklearn import svm

>>> iris = datasets.load_iris()
>>> iris.data.shape, iris.target.shape
((150, 4), (150,))
```

接下来可以快速采样到原数据集的 40% 作为测试集，从而测试 (评估) 我们的分类器：

```
>>> X_train, X_test, y_train, y_test = train_test_split(
...     iris.data, iris.target, test_size=0.4, random_state=0)

>>> X_train.shape, y_train.shape
((90, 4), (90,))
>>> X_test.shape, y_test.shape
((60, 4), (60,))

>>> clf = svm.SVC(kernel='linear', C=1).fit(X_train, y_train)
>>> clf.score(X_test, y_test)
0.96...
```

当设置评估器的不同设置 [hyperparameters（超参数）] 参数时，例如，手动为 SVM 设置 C 参数，由于在训练集上，可通过调整参数 C 设置使估计器的性能达到最佳状态。但是在测试集上可能会出现过拟合的情况。此时，测试集上的信息反馈足以颠覆训练好的模型，评估的指标不再有效反映出模型的泛化性能。为了解决此类问题，还应该准备另一部分被称为验证集（validation set）的数据集，这样当模型训练完成以后在验证集上对模型进行评估。当验证集上的评估实验比较成功时，在测试集上进行最后的评估。

然而，将原始数据分为 3 个数据集的做法，会大大减少可用于模型学习的样本数量，并且得到的结果依赖于集合对（训练，验证）的随机选择。这个问题可以通过交叉验证（CV）来解决。交叉验证仍需要测试集做最后的模型评估，但不再需要验证集。其中最基本的方法被称为：K-Fold（k 折）交叉验证。k 折交叉验证将训练集划分为 k 个较小的集合（其他方法会在下面描述，主要原则基本相同）。每一个 k 折都会遵循下面的过程。

（1）将 k-1 份训练集子集作为训练集训练模型。

（2）将剩余的 1 份训练集子集作为验证集用于模型验证（也就是利用该数据集计算模型的性能指标，例如准确率）。

k 折交叉验证得出的性能指标是循环计算中每个值的平均值。虽然该方法的计算代价很高，但是它不会浪费太多的数据（如固定任意测试集的情况一样），这一点在处理样本数据集较少的问题（例如，逆向推理）时比较有优势。

5.1.1 计算交叉验证的指标

在 Scikit-Learn 应用中，使用交叉验证最简单的方法是在估计器和数据集上调用 cross_val_score 辅助函数。例如下面的演示代码，功能是通过分割数据拟合模型和计算连续 5 次的分数（每次不同分割），来估计 Linear Kernel 支持向量机在 iris 数据集上的精度。

```
>>> from sklearn.model_selection import cross_val_score
>>> clf = svm.SVC(kernel='linear', C=1)
>>> scores = cross_val_score(clf, iris.data, iris.target, cv=5)
>>> scores
array([ 0.96...,  1.  ...,  0.96...,  0.96...,  1.        ])
```

评分估计的平均得分和 95% 置信区间会输出：

```
>>> print("Accuracy: %0.2f (+/- %0.2f)" % (scores.mean(), scores.std() * 2))
Accuracy: 0.98 (+/- 0.03)
```

在默认情况下，每个 CV 迭代计算的分数是估计器 score 方法的运行结果，可以通过使用参数 scoring 来改变计算方式：

```
>>> from sklearn import metrics
>>> scores = cross_val_score(
...     clf, iris.data, iris.target, cv=5, scoring='f1_macro')
>>> scores
array([ 0.96...,  1\.  ...,  0.96...,  0.96...,  1\.        ])
```

在使用 iris 数据集的情形下，样本在各个目标类别之间是平衡的，因此得到的准确度和 F1-score 几乎相等。当 CV 参数是一个整数时，cross_val_score 默认使用 K-Fold 或 StratifiedK-Fold 策略，后者会在估计器派生自 ClassifierMixin 时使用。也可以通过传入一个交叉验证迭代器来使用其他交叉验证策略，比如：

```
>>> from sklearn.model_selection import ShuffleSplit
>>> n_samples = iris.data.shape[0]
>>> cv = ShuffleSplit(n_splits=3, test_size=0.3, random_state=0)
>>> cross_val_score(clf, iris.data, iris.target, cv=cv)
...
array([ 0.97...,  0.97...,  1\.        ])
```

正如在训练集中保留的数据上测试一个预测器（predictor）是很重要的一样，也应该从训练集中学习预处理（如标准化、特征选择等）和类似的数据转换，并进行预测：

```
>>> from sklearn import preprocessing
>>> X_train, X_test, y_train, y_test = train_test_split(
...     iris.data, iris.target, test_size=0.4, random_state=0)
>>> scaler = preprocessing.StandardScaler().fit(X_train)
>>> X_train_transformed = scaler.transform(X_train)
>>> clf = svm.SVC(C=1).fit(X_train_transformed, y_train)
>>> X_test_transformed = scaler.transform(X_test)
>>> clf.score(X_test_transformed, y_test)
0.9333...
```

Pipeline 可以更加容易地组合估计器，例如，下面是在交叉验证下的代码：

```
>>> from sklearn.pipeline import make_pipeline
>>> clf = make_pipeline(preprocessing.StandardScaler(), svm.SVC(C=1))
>>> cross_val_score(clf, iris.data, iris.target, cv=cv)
...
array([ 0.97...,  0.93...,  0.95...])
```

1. 函数 cross_validate 和多度量评估

在 Scikit-Learn 应用中，函数 cross_validate 与函数 cross_val_score 的区别是它允许指定多个指标进行评估。除了测试得分之外，它还会返回一个包含训练得分、拟合次数、得分次数（score-times）的一个字典。对于单个度量评估来说，其中参数 scoring 是一个字符串，可以为 None，参数 keys 将是 - ['test_score', 'fit_time', 'score_time']。

而对于多度量评估来说，返回值是一个带有以下 keys 的字典：['test_<scorer1_name>', 'test_<scorer2_name>', 'test_<scorer...>', 'fit_time', 'score_time']。

参数 return_train_score 的默认值为 True，它增加了所有得分器（scorers）的训练得分 keys。如果不需要训练 scores，则应将其明确设置为 False。

在使用函数 cross_validate 时，可以将多个指标指定为 predefined scorer names（预定义的得分器的名称）的 list、tuple 或者 set 形式。例如：

```
>>> from sklearn.model_selection import cross_validate
>>> from sklearn.metrics import recall_score
>>> scoring = ['precision_macro', 'recall_macro']
>>> clf = svm.SVC(kernel='linear', C=1, random_state=0)
>>> scores = cross_validate(clf, iris.data, iris.target, scoring=scoring,
...                         cv=5, return_train_score=False)
>>> sorted(scores.keys())
['fit_time', 'score_time', 'test_precision_macro', 'test_recall_macro']
>>> scores['test_recall_macro']
array([ 0.96...,  1\. ...,  0.96...,  0.96...,  1\.        ])
```

也可以将指标作为一个字典 mapping 的预定义得分器名称或自定义的得分函数，例如：

```
>>> from sklearn.metrics.scorer import make_scorer
>>> scoring = {'prec_macro': 'precision_macro',
...            'rec_micro': make_scorer(recall_score, average='macro')}
>>> scores = cross_validate(clf, iris.data, iris.target, scoring=scoring,
```

```
...                       cv=5, return_train_score=True)
>>> sorted(scores.keys())
['fit_time', 'score_time', 'test_prec_macro', 'test_rec_micro',
 'train_prec_macro', 'train_rec_micro']
>>> scores['train_rec_micro']
array([ 0.97...,  0.97...,  0.99...,  0.98...,  0.98...])
```

例如,下面是一个使用单一指标的 cross_validate 的例子:

```
>>> scores = cross_validate(clf, iris.data, iris.target,
...                          scoring='precision_macro')
>>> sorted(scores.keys())
['fit_time', 'score_time', 'test_score', 'train_score']
```

2. 通过交叉验证获取预测

除了返回值的结果不同之外,函数 cross_val_predict 具有和函数 cross_val_score 相同的接口。对于每一个输入的元素来说,如果其在测试集合中,将会得到预测结果。交叉验证策略会有且仅有一次可用的元素提交到测试集合,多次会抛出一个异常。这些预测可以用于评价分类器的效果,例如:

```
>>> from sklearn.model_selection import cross_val_predict
>>> predicted = cross_val_predict(clf, iris.data, iris.target, cv=10)
>>> metrics.accuracy_score(iris.target, predicted)
0.973...
```

⚠️ **注意** 上述计算的结果和 cross_val_score 有轻微的差别,因为后者用另一种方式组织元素。

请看下面的实例文件 jiaocha01.py,演示了嵌套 CV 与非嵌套 CV 的实现过程。本实例分别对鸢尾花数据集分类器进行了非嵌套 CV 和嵌套 CV 策略处理。嵌套 CV 通常用于训练超参数也需要优化的模型,能够估计基础模型及其(超)参数搜索的泛化误差。选择使用最大化非嵌套 CV 的参数,会使模型偏向于数据集,从而产生过于乐观的分数。在本实例中,使用具有非线性内核的支持向量分类器通过网格搜索构建具有优化超参数的模型,通过计算非嵌套 CV 和嵌套 CV 策略的分数之间的差异来比较它们的性能。

```
#随机试验次数
NUM_TRIALS = 30

# Load the dataset
iris = load_iris()
X_iris = iris.data
y_iris = iris.target

#设置可能的参数值以优化
p_grid = {"C": [1, 10, 100],
          "gamma": [.01, .1]}

#将使用带有"rbf"核的支持向量分类器
svm = SVC(kernel="rbf")

#存储分数的数组
```

```python
non_nested_scores = np.zeros(NUM_TRIALS)
nested_scores = np.zeros(NUM_TRIALS)

#循环训练
for i in range(NUM_TRIALS):

    #为内部和外部循环选择交叉验证技术,独立于数据集。例如GroupKFold、LeaveOneOut、
     LeaveOneGroupOut等
    inner_cv = KFold(n_splits=4, shuffle=True, random_state=i)
    outer_cv = KFold(n_splits=4, shuffle=True, random_state=i)

    #非嵌套参数搜索与评分
    clf = GridSearchCV(estimator=svm, param_grid=p_grid, cv=inner_cv)
    clf.fit(X_iris, y_iris)
    non_nested_scores[i] = clf.best_score_

    #参数优化的嵌套CV
    nested_score = cross_val_score(clf, X=X_iris, y=y_iris, cv=outer_cv)
    nested_scores[i] = nested_score.mean()

score_difference = non_nested_scores - nested_scores

print("Average difference of {:6f} with std. dev. of {:6f}."
      .format(score_difference.mean(), score_difference.std()))

#绘制嵌套CV和非嵌套CV每次试验的得分
plt.figure()
plt.subplot(211)
non_nested_scores_line, = plt.plot(non_nested_scores, color='r')
nested_line, = plt.plot(nested_scores, color='b')
plt.ylabel("score", fontsize="14")
plt.legend([non_nested_scores_line, nested_line],
           ["Non-Nested CV", "Nested CV"],
           bbox_to_anchor=(0, .4, .5, 0))
plt.title("Non-Nested and Nested Cross Validation on Iris Dataset",
          x=.5, y=1.1, fontsize="15")

#绘制差异条形图
plt.subplot(212)
difference_plot = plt.bar(range(NUM_TRIALS), score_difference)
plt.xlabel("Individual Trial #")
plt.legend([difference_plot],
           ["Non-Nested CV - Nested CV Score"],
           bbox_to_anchor=(0, 1, .8, 0))
plt.ylabel("score difference", fontsize="14")

plt.show()
```

jiaocha01.py 的执行效果如图 5-1 所示。

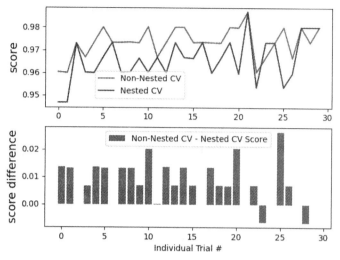

图 5-1 jiaocha01.py 的执行效果

没有嵌套 CV 的模型选择使用相同的数据来调整模型参数和评估模型性能，因此，信息可能会"泄露"到模型中并过度拟合数据。这种影响的大小主要取决于数据集的大小和模型的稳定性。为了避免这个问题，嵌套 CV 有效地使用了一系列"训练/验证/测试集"拆分功能。在内部循环中（由 GridSearchCV 执行的），通过将模型拟合到每个训练集来近似最大化分数，然后在验证集上选择（超）参数时直接最大化。在外部循环中（此处为 cross_val_score），通过对多个数据集拆分的测试集分数进行平均来估计泛化误差。

5.1.2 交叉验证迭代器

假设一些数据是独立的和相同分布的（i.i.d），假定所有的样本来源于相同的生成过程，并假设生成过程没有记忆过去生成的样本，在这种情况下可以使用下面的交叉验证器。

⚠️ 注 意　i.i.d 数据是机器学习理论中的一个常见假设，在实践中很少成立。如果知道样本是使用时间相关的过程生成的，则使用 time-series aware cross-validation scheme 更安全。同样，如果我们知道生成过程具有群体结构（group structure）[从不同主体（subjects）、实验（experiments）、测量设备（measurement devices）收集的样本]，则使用 group-wise cross-validation 更安全。

1. K-Fold（k 折）

K-Fold 将所有的样例划分为 k 个组，称为折叠（fold）[如果 $k = n$，这等价于留一（Leave One Out）策略]，都具有相同的大小（如果可能）。预测函数学习时使用 $k-1$ 个折叠中的数据，最后一个剩下的折叠会用于测试。

下面是一个在 4 个样例的数据集上使用 2-fold 进行交叉验证的例子：

```
>>> import numpy as np
>>> from sklearn.model_selection import KFold

>>> X = ["a", "b", "c", "d"]
>>> kf = KFold(n_splits=2)
>>> for train, test in kf.split(X):
...     print("%s  %s" % (train, test))
[2 3] [0 1]
[0 1] [2 3]
```

每个折叠由两个数组（arrays）组成，一个作为训练集合（training set），另一个作为测试集合（test set）。此时可以通过使用 numpy 的索引创建"训练/测试"集合：

```
>>> X = np.array([[0., 0.], [1., 1.], [-1., -1.], [2., 2.]])
>>> y = np.array([0, 1, 0, 1])
>>> X_train, X_test, y_train, y_test = X[train], X[test], y[train], y[test]
```

2. 重复 K-Flod 交叉验证

在 Scikit-Learn 应用中，RepeatedKFold 用于重复运行 n 次 K-Fold，当需要运行时可以 n 次使用 K-Fold，在每次重复中产生不同的分割。例如，下面是 2 折 K-Fold 重复 2 次的例子：

```
>>> import numpy as np
>>> from sklearn.model_selection import RepeatedKFold
>>> X = np.array([[1, 2], [3, 4], [1, 2], [3, 4]])
>>> random_state = 12883823
>>> rkf = RepeatedKFold(n_splits=2, n_repeats=2, random_state=random_state)
>>> for train, test in rkf.split(X):
...     print("%s  %s" % (train, test))
...
[2 3] [0 1]
[0 1] [2 3]
[0 2] [1 3]
[1 3] [0 2]
```

类似地，RepeatedStratifiedKFold 在每个重复中以不同的随机化重复 n 次分层的 K-Fold。

5.2 调整估计器的超参数

超参数即不直接在估计器内学习的参数。在 Scikit-Learn 应用中，超参数作为估计器类中构造函数的参数进行传递。搜索超参数空间可以获得最好的交叉验证分数，这个方法是值得提倡的。通过这种方式，构造估计器时被提供的任何参数或许都能被优化。具体来说，要获取到给定估计器的所有参数的名称和当前值，可以使用函数 estimator.get_params() 实现。

扫码观看本节视频讲解

搜索包括：
- 估计器（回归器或分类器，例如 sklearn.svm.SVC()）；
- 参数空间；
- 搜寻或采样候选的方法；
- 交叉验证方案；
- 计分函数。

有些模型支持专业化的、高效的参数搜索策略。在 Scikit-Learn 应用中提供了两种采样搜索候选的通用方法。
- 对于给定的值，GridSearchCV 考虑了所有参数组合。
- RandomizedSearchCV 可以从具有指定分布的参数空间中抽取给定数量的候选。

5.2.1 网格追踪法：穷尽的网格搜索

在 Scikit-Learn 应用中，GridSearchCV 提供的网格搜索从通过 param_grid 参数确定的网格参数值中全面生成候选。例如下面的 param_grid：

```
param_grid = [
  {'C': [1, 10, 100, 1000], 'kernel': ['linear']},
  {'C': [1, 10, 100, 1000], 'gamma': [0.001, 0.0001], 'kernel': ['rbf']},
]
```

在上述搜索网格中，一个具有线性内核并且 C 在 [1,10,100,1000] 中取值；另一个具有 RBF 内核，C 值的交叉乘积范围在 [1,10,100,1000]，gamma 在 [0.001,0.0001] 中取值。GridSearchCV 实例实现了常用估计器 API：当在数据集上"拟合"时，参数值的所有可能的组合都会被评估，从而计算出最佳的组合。

请看下面的实例文件 jiaocha02.py，功能是使用带有交叉验证的网格搜索实现参数估计功能。本实例展示了通过交叉验证优化分类器的方法，这是使用 GridSearchCV 开发集上的对象完成的，该对象仅包含可用标记数据的一半。然后在模型选择步骤中未使用的专用评估集上测量所选超参数和训练模型的性能。

```
from sklearn import datasets
from sklearn.model_selection import train_test_split
from sklearn.model_selection import GridSearchCV
from sklearn.metrics import classification_report
from sklearn.svm import SVC

print(__doc__)

#加载数字数据集
digits = datasets.load_digits()

#要对这些数据应用分类器,我们需要将图像展平,将数据转换为(样本、特征)矩阵:
n_samples = len(digits.images)
X = digits.images.reshape(n_samples, -1)
y = digits.target
```

```python
#将数据集分成两个相等的部分
X_train, X_test, y_train, y_test = train_test_split(
    X, y, test_size=0.5, random_state=0)

#通过交叉验证设置参数
tuned_parameters = [{'kernel': ['rbf'], 'gamma': [1e-3, 1e-4],
                     'C': [1, 10, 100, 1000]},
                    {'kernel': ['linear'], 'C': [1, 10, 100, 1000]}]

scores = ['precision', 'recall']

for score in scores:
    print("# Tuning hyper-parameters for %s" % score)
    print()

    clf = GridSearchCV(
        SVC(), tuned_parameters, scoring='%s_macro' % score
    )
    clf.fit(X_train, y_train)

    print("Best parameters set found on development set:")
    print()
    print(clf.best_params_)
    print()
    print("Grid scores on development set:")
    print()
    means = clf.cv_results_['mean_test_score']
    stds = clf.cv_results_['std_test_score']
    for mean, std, params in zip(means, stds, clf.cv_results_['params']):
        print("%0.3f (+/-%0.03f) for %r"
              % (mean, std * 2, params))
    print()

    print("Detailed classification report:")
    print()
    print("The model is trained on the full development set.")
    print("The scores are computed on the full evaluation set.")
    print()
    y_true, y_pred = y_test, clf.predict(X_test)
    print(classification_report(y_true, y_pred))
    print()
```

执行后会输出：

```
Best parameters set found on development set:

{'C': 10, 'gamma': 0.001, 'kernel': 'rbf'}
```

```
Grid scores on development set:

0.986 (+/-0.016) for {'C': 1, 'gamma': 0.001, 'kernel': 'rbf'}
0.959 (+/-0.028) for {'C': 1, 'gamma': 0.0001, 'kernel': 'rbf'}
0.988 (+/-0.017) for {'C': 10, 'gamma': 0.001, 'kernel': 'rbf'}
0.982 (+/-0.026) for {'C': 10, 'gamma': 0.0001, 'kernel': 'rbf'}
0.988 (+/-0.017) for {'C': 100, 'gamma': 0.001, 'kernel': 'rbf'}
0.983 (+/-0.026) for {'C': 100, 'gamma': 0.0001, 'kernel': 'rbf'}
0.988 (+/-0.017) for {'C': 1000, 'gamma': 0.001, 'kernel': 'rbf'}
0.983 (+/-0.026) for {'C': 1000, 'gamma': 0.0001, 'kernel': 'rbf'}
0.974 (+/-0.012) for {'C': 1, 'kernel': 'linear'}
0.974 (+/-0.012) for {'C': 10, 'kernel': 'linear'}
0.974 (+/-0.012) for {'C': 100, 'kernel': 'linear'}
0.974 (+/-0.012) for {'C': 1000, 'kernel': 'linear'}

Detailed classification report:

The model is trained on the full development set.
The scores are computed on the full evaluation set.

              precision    recall  f1-score   support

           0       1.00      1.00      1.00        89
           1       0.97      1.00      0.98        90
           2       0.99      0.98      0.98        92
           3       1.00      0.99      0.99        93
           4       1.00      1.00      1.00        76
           5       0.99      0.98      0.99       108
           6       0.99      1.00      0.99        89
           7       0.99      1.00      0.99        78
           8       1.00      0.98      0.99        92
           9       0.99      0.99      0.99        92

    accuracy                           0.99       899
   macro avg       0.99      0.99      0.99       899
weighted avg       0.99      0.99      0.99       899


# Tuning hyper-parameters for recall

Best parameters set found on development set:

{'C': 10, 'gamma': 0.001, 'kernel': 'rbf'}

Grid scores on development set:

0.986 (+/-0.019) for {'C': 1, 'gamma': 0.001, 'kernel': 'rbf'}
0.957 (+/-0.028) for {'C': 1, 'gamma': 0.0001, 'kernel': 'rbf'}
0.987 (+/-0.019) for {'C': 10, 'gamma': 0.001, 'kernel': 'rbf'}
```

```
0.981 (+/-0.028) for {'C': 10, 'gamma': 0.0001, 'kernel': 'rbf'}
0.987 (+/-0.019) for {'C': 100, 'gamma': 0.001, 'kernel': 'rbf'}
0.982 (+/-0.026) for {'C': 100, 'gamma': 0.0001, 'kernel': 'rbf'}
0.987 (+/-0.019) for {'C': 1000, 'gamma': 0.001, 'kernel': 'rbf'}
0.982 (+/-0.026) for {'C': 1000, 'gamma': 0.0001, 'kernel': 'rbf'}
0.971 (+/-0.010) for {'C': 1, 'kernel': 'linear'}
0.971 (+/-0.010) for {'C': 10, 'kernel': 'linear'}
0.971 (+/-0.010) for {'C': 100, 'kernel': 'linear'}
0.971 (+/-0.010) for {'C': 1000, 'kernel': 'linear'}

Detailed classification report:

The model is trained on the full development set.
The scores are computed on the full evaluation set.

              precision    recall  f1-score   support

           0       1.00      1.00      1.00        89
           1       0.97      1.00      0.98        90
           2       0.99      0.98      0.98        92
           3       1.00      0.99      0.99        93
           4       1.00      1.00      1.00        76
           5       0.99      0.98      0.99       108
           6       0.99      1.00      0.99        89
           7       0.99      1.00      0.99        78
           8       1.00      0.98      0.99        92
           9       0.99      0.99      0.99        92

    accuracy                           0.99       899
   macro avg       0.99      0.99      0.99       899
weighted avg       0.99      0.99      0.99       899
```

5.2.2 随机参数优化

尽管使用参数设置网格法是目前最广泛使用的参数优化方法，其他搜索方法也具有更有利的性能。在 Scikit-Learn 应用中，RandomizedSearchCV 实现了对参数的随机搜索，其中每个设置都是从可能的参数值的分布中进行取样。这对于穷举搜索来说有两个主要优势：
- 可以选择独立于参数个数和可能值的预算。
- 添加不影响性能的参数不会降低效率。

指定如何取样的参数是使用字典完成的，类似于为 GridSearchCV 指定参数。此外，通过参数 n_iter 设置迭代次数，即取样候选项数或取样的迭代次数。对于每个参数来说，可以指定在可能值上的分布或离散选择的列表（均匀取样）：

```
{'C': scipy.stats.expon(scale=100), 'gamma': scipy.stats.expon(scale=.1),
  'kernel': ['rbf'], 'class_weight':['balanced', None]}
```

上述代码使用 scipy.stats 模块，它包含许多用于采样参数的有用分布，如 expon、gamma、uniform 或 randint。原则上，任何函数都可以通过提供一个随机变量样本（rvs）方法来采样一个值。对 rvs 函数的调用应在连续调用中提供来自可能参数值的独立随机样本。

请看下面的实例文件 jiaocha03.py，功能是比较随机搜索和网格搜索的使用和效率。

```python
#获取一些数据
X, y = load_digits(return_X_y=True)

#构建分类器
clf = SGDClassifier(loss='hinge', penalty='elasticnet',
                    fit_intercept=True)


#报告最佳分数的实用函数
def report(results, n_top=3):
    for i in range(1, n_top + 1):
        candidates = np.flatnonzero(results['rank_test_score'] == i)
        for candidate in candidates:
            print("Model with rank: {0}".format(i))
            print("Mean validation score: {0:.3f} (std: {1:.3f})"
                  .format(results['mean_test_score'][candidate],
                          results['std_test_score'][candidate]))
            print("Parameters: {0}".format(results['params'][candidate]))
            print("")

#指定要从中采样的参数和分布
param_dist = {'average': [True, False],
              'l1_ratio': stats.uniform(0, 1),
              'alpha': loguniform(1e-4, 1e0)}

#运行随机搜索
n_iter_search = 20
random_search = RandomizedSearchCV(clf, param_distributions=param_dist,
                                   n_iter=n_iter_search)

start = time()
random_search.fit(X, y)
print("RandomizedSearchCV took %.2f seconds for %d candidates"
      " parameter settings." % ((time() - start), n_iter_search))
report(random_search.cv_results_)

#对所有参数使用完整网格
param_grid = {'average': [True, False],
              'l1_ratio': np.linspace(0, 1, num=10),
              'alpha': np.power(10, np.arange(-4, 1, dtype=float))}

#运行网格搜索
grid_search = GridSearchCV(clf, param_grid=param_grid)
```

```
start = time()
grid_search.fit(X, y)

print("GridSearchCV took %.2f seconds for %d candidate parameter settings."
      % (time() - start, len(grid_search.cv_results_['params'])))
report(grid_search.cv_results_)
```

通过上述代码，比较了随机搜索和网格搜索以优化线性 SVM 的超参数和 SGD 训练，同时搜索所有影响学习的参数（估计量的数量除外，这会造成时间／质量的权衡）。随机搜索和网格搜索探索完全相同的参数空间。参数设置的结果非常相似，而随机搜索的运行时间却大大降低。执行本实例后会输出：

```
RandomizedSearchCV took 31.72 seconds for 20 candidates parameter settings.
Model with rank: 1
Mean validation score: 0.920 (std: 0.028)
Parameters: {'alpha': 0.07316411520495676, 'average': False, 'l1_ratio': 0.29007760721044407}

Model with rank: 2
Mean validation score: 0.920 (std: 0.029)
Parameters: {'alpha': 0.0005223493320259539, 'average': True, 'l1_ratio': 0.7936977033574206}

Model with rank: 3
Mean validation score: 0.918 (std: 0.031)
Parameters: {'alpha': 0.00025790124268693137, 'average': True, 'l1_ratio': 0.5699649107012649}

GridSearchCV took 166.67 seconds for 100 candidate parameter settings.
Model with rank: 1
Mean validation score: 0.931 (std: 0.026)
Parameters: {'alpha': 0.0001, 'average': True, 'l1_ratio': 0.0}

Model with rank: 2
Mean validation score: 0.928 (std: 0.030)
Parameters: {'alpha': 0.0001, 'average': True, 'l1_ratio': 0.1111111111111111}

Model with rank: 3
Mean validation score: 0.927 (std: 0.026)
Parameters: {'alpha': 0.0001, 'average': True, 'l1_ratio': 0.5555555555555556}
```

随机搜索的性能可能稍差，并且可能是受噪声影响，不会延续到保留的测试集。注意，在实践应用中，人们不会使用网格搜索同时搜索这么多不同的参数，而是只选择最重要的参数。

5.3 模型评估：量化预测的质量

在 Scikit-Learn 应用中，有如下 3 种用于评估模型预测质量的 API。
- 估计器得分的方法（Estimator score method）：在估计器（Estimators）中有一个得分方法，为其解决的问题提供了默认的评估标准。

扫码观看本节视频讲解

- 评分参数（Scoring parameter）：模型评估工具（Model-evaluation tools）使用交叉验证（cross-validation）（如 model_selection.cross_val_score 和 model_selection.GridSearchCV）依靠内部评分策略实现。
- 指标函数（Metric functions）：metrics 模块实现了针对特定目的评估预测误差的函数。

5.3.1 得分参数 scoring：定义模型评估规则

在 Scikit-Learn 应用中，得分参数 scoring 被用在模型选择（Model selection）和评估（evaluation）中，例如 model_selection.GridSearchCV 和 model_selection.cross_val_score，采用参数 scoring 来控制它们对评估的估计量（estimators evaluated）应用的指标。

1. 得分参数 scoring 的应用场景：预定义值

在现实应用中，可以使用参数 scoring 指定一个记分对象（scorer object）。所有记分对象遵循惯例"较高的返回值优于较低的返回值"。因此，测量模型和数据之间距离的度量（metrics），如 metrics.mean_squared_error 可用作返回指数（metric）的否定值（negated value）的 neg_mean_squared_error。例如，下面的代码演示了使用得分参数 scoring 预定义值的用法：

```
>>> from sklearn import svm, datasets
>>> from sklearn.model_selection import cross_val_score
>>> iris = datasets.load_iris()
>>> X, y = iris.data, iris.target
>>> clf = svm.SVC(probability=True, random_state=0)
>>> cross_val_score(clf, X, y, scoring='neg_log_loss')
array([-0.07..., -0.16..., -0.06...])
>>> model = svm.SVC()
>>> cross_val_score(model, X, y, scoring='wrong_choice')
Traceback (most recent call last):
ValueError: 'wrong_choice' is not a valid scoring value. Valid options are ['accuracy', 'adjusted_mutual_info_score', 'adjusted_rand_score', 'average_precision', 'completeness_score', 'explained_variance', 'f1', 'f1_macro', 'f1_micro', 'f1_samples', 'f1_weighted', 'fowlkes_mallows_score', 'homogeneity_score', 'mutual_info_score', 'neg_log_loss', 'neg_mean_absolute_error', 'neg_mean_squared_error', 'neg_mean_squared_log_error', 'neg_median_absolute_error', 'normalized_mutual_info_score', 'precision', 'precision_macro', 'precision_micro', 'precision_samples', 'precision_weighted', 'r2', 'recall', 'recall_macro', 'recall_micro', 'recall_samples', 'recall_weighted', 'roc_auc', 'v_measure_score']
```

2. 得分参数 scoring 的应用场景：根据 metric 函数定义评分策略

在 Scikit-Learn 应用中，模块 sklearn.metrics 还公开了一组测量预测误差的简单函数，在其中给出了基础真实的数据和预测。

- 函数以_score结尾返回一个值来实现最大化，越高越好。
- 函数以_error 或_loss结尾返回一个值实现最小化，越低越好。当使用 make_scorer 转换成记分对象（scorer object）时，将参数greater_is_better设置为 False（默认为 True）。

许多指标（metrics）没有被用作得分（scoring）值的名称，有时是因为它们需要额外的参数，例如 fbeta_score。在这种情况下，您需要生成一个适当的评分对象（scoring object）。生成可评估对象进行评分（callable object for scoring）的最简单方法是使用函数 make_scorer，该函数将指数（metrics）转换为可调用的模型评估（model evaluation）。

第一个例子是从库中包含一个非默认值参数的现有指数函数（existing metric function），例如，fbeta_score 函数的 beta 参数：

```
>>> from sklearn.metrics import fbeta_score, make_scorer
>>> ftwo_scorer = make_scorer(fbeta_score, beta=2)
>>> from sklearn.model_selection import GridSearchCV
>>> from sklearn.svm import LinearSVC
>>> grid = GridSearchCV(LinearSVC(), param_grid={'C': [1, 10]}, scoring=ftwo_scorer)
```

第二个例子是使用 make_scorer 从简单的 Python 函数构建一个完全自定义的评分对象（custom scorer object）。下面是建立自定义评分对象的示例，并使用了参数 greater_is_better。

```
>>> import numpy as np
>>> def my_custom_loss_func(ground_truth, predictions):
...     diff = np.abs(ground_truth - predictions).max()
...     return np.log(1 + diff)
...
>>> # loss_func will negate the return value of my_custom_loss_func,
>>> #  which will be np.log(2), 0.693, given the values for ground_truth
>>> #  and predictions defined below.
>>> loss  = make_scorer(my_custom_loss_func, greater_is_better=False)
>>> score = make_scorer(my_custom_loss_func, greater_is_better=True)
>>> ground_truth = [[1], [1]]
>>> predictions  = [0, 1]
>>> from sklearn.dummy import DummyClassifier
>>> clf = DummyClassifier(strategy='most_frequent', random_state=0)
>>> clf = clf.fit(ground_truth, predictions)
>>> loss(clf,ground_truth, predictions)
-0.69...
>>> score(clf,ground_truth, predictions)
0.69...
```

3. 得分参数 scoring 的应用场景：实现自己的评分对象

在 Scikit-Learn 应用中，可以从头开始构建自己的评分对象，而不使用 make_scorer 来生成更加灵活的模型评分对象。在实现自己的评分对象时，需要符合以下两个规则所指定的协议。

- 可以使用参数（estimator、X、y）来调用它，其中，estimator 是要被评估的模型，X 是验证数据，y 是 X（在有监督的情况下）或 None（在无监督的情况下）已经被标注的真实数据目标。
- 返回一个浮点数，用于对 X 进行量化 estimator 的预测质量，参考 y。再次，按照惯例，因为更高的数字的效果会更好，所以如果你的 scorer 返回 loss，那么应该否定这个值，因为这样的效果不是很好。

5.3.2 分类指标

在 Scikit-Learn 应用中，模块 sklearn.metrics 实现了用 loss、score 和 utility 函数来衡量分类（classification）的性能。某些指标（metrics）可能需要正类（positive class）、置信度值（confidence values）或二进制决策值（binary decisions values）的概率估计。大多数的实现允许每个样本通过 sample_weight 参数为总分（overall score）提供加权贡献（weighted contribution）。

1. 从二分到多分类和 multilabel

很多 metrics 是为二分类任务（Binary Classification Tasks）定义的（例如 f1_score、roc_auc_score）。在这些情况下，默认仅评估 positive label（正标签）。假设在默认情况下，正类（positive label）标记为 1（尽管可以通过 pos_label 参数进行配置）。

当将二分指标（Binary Metric）扩展为多类（multiclass）或多标签（multilabel）问题时，数据将被视为二分问题的集合，每个类都有一个。然后可以使用多种方法在整个类中计算平均 Binary Metrics，每种类在某些情况下可能会有用。如果可用，应该使用参数 average 来选择它们。

- macro（宏）：用于简单地计算 Binary Metrics 的平均值，赋予每个类别相同的权重。在处理不常见的类的问题上，宏观平均（macro-averaging）可能是突出表现的一种手段。另外，对所有类别做同样重要的假设通常是不客观的，因此 macro-averaging 将过度强调不频繁类的典型的低性能。
- 加权（weighted）：功能是通过计算其在真实数据样本中的存在来对每个类的 score 进行加权的 Binary Metrics 的平均值来计算类不平衡。
- micro（微）：功能是给每个样本类对（sample-class pair）对总体指数（overall metric）（sample-class 权重的结果除外）等同的贡献。除了对每个类别的 metric 进行求和之外，这个和构成每个类计算一个商的除数和约数。在多标签设置（multilabel settings）中，Micro-averaging 可能是优先选择项，包括要忽略多数类（majority class）的多类分类（multiclass classification）。
- 样本（samples）：仅适用于多标签问题（multilabel problems），它不计算每个类别的值，而是计算评估数据（evaluation data）中的每个样本的真实和预测类别（true and predicted classes）的指标（metric），并返回加权平均（sample_weight-weighted）。
- 设置 average=None，将返回一个 array 与每个类的 score。

虽然将多类数据（multiclass data）提供给 metric，例如，将二分类目标（binary targets）作为类标签的数组，需要将多标签数据指定为指示符矩阵，其中，cell [i, j] 包含值 1，则样本 i 具有标号 j，否则为值 0。

2. 精确度得分

在 Scikit-Learn 应用中，函数 accuracy_score 用于计算 accuracy 精确度，默认值是返回预测的分数（默认），也可以返回计数（normalize=False）。

在多标签分类中，函数 accuracy_score 返回子集精度。如果样本的整套预测标签与真正的标签组合匹配，则子集精度为 1.0，否则为 0.0。

如果 $\hat{y}i$ 是第 i 个样本的预测值，y_i 是相应的真实值，则 $n_{samples}$ 上的正确预测的分数被定义为：

$$\text{accuracy}(y, \hat{y}) = \frac{1}{n_{\text{samples}}} \sum_{i=0}^{n_{\text{samples}}-1} 1(\hat{y}i = yi)$$

其中，1(x)是指示函数，例如下面的演示代码：

```
>>> import numpy as np
>>> from sklearn.metrics import accuracy_score
>>> y_pred = [0, 2, 1, 3]
>>> y_true = [0, 1, 2, 3]
>>> accuracy_score(y_true, y_pred)
0.5
>>> accuracy_score(y_true, y_pred, normalize=False)
2
```

下面是在具有二分标签指示符的多标签情况下的使用情况：

```
>>> accuracy_score(np.array([[0, 1], [1, 1]]), np.ones((2, 2)))
0.5
```

3. 混淆矩阵

在 Scikit-Learn 应用中，函数 confusion_matrix 通过计算 confusion matrix（混淆矩阵）来评估分类的准确性。根据定义，混淆矩阵中的 entry（条目）i 和 j，实际上在 group i 中的 observations（观察数），预测在 group j 中。例如下面的演示代码：

```
>>> from sklearn.metrics import confusion_matrix
>>> y_true = [2, 0, 2, 2, 0, 1]
>>> y_pred = [0, 0, 2, 2, 0, 2]
>>> confusion_matrix(y_true, y_pred)
array([[2, 0, 0],
 [0, 0, 1],
 [1, 0, 2]])
```

对于二分类问题（Binary Problems）来说，可以得到真 negatives、假 positives，假 negatives 和真 positives 的数量如下：

```
>>> y_true = [0, 0, 0, 1, 1, 1, 1, 1]
>>> y_pred = [0, 1, 0, 1, 0, 1, 0, 1]
>>> tn, fp, fn, tp = confusion_matrix(y_true, y_pred).ravel()
>>> tn, fp, fn, tp
(2, 1, 2, 3)
```

请看下面的实例文件 jiaocha02.py，功能是使用混淆矩阵来评估分类器（classifier）的输出质量。

```
import numpy as np
import matplotlib.pyplot as plt

from sklearn import svm, datasets
from sklearn.model_selection import train_test_split
from sklearn.metrics import plot_confusion_matrix

#导入一些数据
iris = datasets.load_iris()
```

```python
X = iris.data
y = iris.target
class_names = iris.target_names

#将数据分为训练集和测试集
X_train, X_test, y_train, y_test = train_test_split(X, y, random_state=0)

#运行分类器,使用一个过于正则化(C太低)的模型来查看对结果的影响
classifier = svm.SVC(kernel='linear', C=0.01).fit(X_train, y_train)

np.set_printoptions(precision=2)

#绘制非标准化混淆矩阵
titles_options = [("Confusion matrix, without normalization", None),
                  ("Normalized confusion matrix", 'true')]
for title, normalize in titles_options:
    disp = plot_confusion_matrix(classifier, X_test, y_test,
                                 display_labels=class_names,
                                 cmap=plt.cm.Blues,
                                 normalize=normalize)
    disp.ax_.set_title(title)
    print(title)
    print(disp.confusion_matrix)

plt.show()
```

jiaocha02.py 的执行效果如图 5-2 所示。

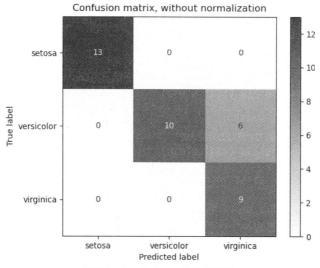

图 5-2　jiaocha02.py 的执行效果

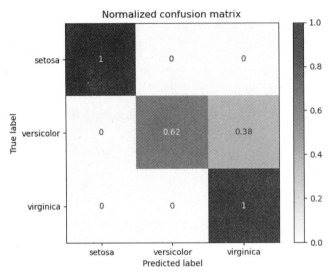

图 5-2　jiaocha02.py 的执行效果（续）

在图 5-2 的执行效果中，对角线元素表示预测标签等于真实标签的点数，而非对角线元素是那些被分类器错误标记的点。混淆矩阵的对角线值越高越好，表明许多正确的预测。这些效果图显示了按类别支持大小（每个类别中的元素数）进行归一化和未归一化的混淆矩阵。在类不平衡的情况下，这种归一化可以更直观地解释哪个类被错误分类。

本实例的执行结果并不理想，因为我们对正则化参数 C 的选择不是最好的。在实际应用中，通常使用调整估计器的超参数来选择此参数。

第 6 章

数据集转换

　　Scikit-Learn 提供了一个用于实现数据集转换功能的库,它也许会清理(clean)、减少(reduce)、扩展(expand)或生成(generate)特征表示。在本章的内容中,将详细介绍使用 Scikit-Learn 实现数据集转换的知识,为读者步入本书后面知识的学习打下基础。

6.1 Pipeline（管道）和 FeatureUnion（特征联合）

Transformer 通常与分类器、回归器或其他估计器相结合，以构建出功能强大的复合估计器。构建复合估计器最常见的工具是 Pipeline，Pipeline 通常与 FeatureUnion 结合使用，FeatureUnion 将转换器的输出连接到复合特征空间中。

扫码观看本节视频讲解

6.1.1 Pipeline：链式评估器

Pipeline 可以把多个评估器链接成一个，这个功能是很有用的，因为处理数据的步骤一般都是固定的，例如特征选择、标准化和分类。Pipeline 主要有两个目的——便捷性和封装性，只要对数据调用 fit 和 predict 一次即可适配所有的一系列评估器。安全性训练转换器和预测器使用的是相同样本，管道有助于防止来自测试数据的统计数据泄露到交叉验证的训练模型中。管道中的所有评估器，除了最后一个评估器，都必须是转换器。

在 Scikit-Learn 应用中，使用一系列键值对 (key, value) 来构建管道，其中 key 是给这个步骤起的名字，value 是一个评估器对象，例如：

```
>>> from sklearn.pipeline import Pipeline
>>> from sklearn.svm import SVC
>>> from sklearn.decomposition import PCA
>>> estimators = [('reduce_dim', PCA()), ('clf', SVC())]
>>> pipe = Pipeline(estimators)
>>> pipe
Pipeline(memory=None,
     steps=[('reduce_dim', PCA(copy=True,...)),
            ('clf', SVC(C=1.0,...))])
```

功能函数 make_pipeline() 是构建管道的缩写，功能是接收多个评估器并返回一个管道，且自动填充评估器名，例如：

```
>>> from sklearn.pipeline import make_pipeline
>>> from sklearn.naive_bayes import MultinomialNB
>>> from sklearn.preprocessing import Binarizer
>>> make_pipeline(Binarizer(), MultinomialNB())
Pipeline(memory=None,
     steps=[('binarizer', Binarizer(copy=True, threshold=0.0)),
            ('multinomialnb', MultinomialNB(alpha=1.0,
                                    class_prior=None,
                                    fit_prior=True))])
```

管道中的评估器作为一个列表保存在 steps 属性内：

```
>>> pipe.steps[0]
('reduce_dim', PCA(copy=True, iterated_power='auto', n_components=None,
```

```
random_state=None,
    svd_solver='auto', tol=0.0, whiten=False))
```

并作为 dict（字典）保存在 named_steps 中：

```
>>> pipe.named_steps['reduce_dim']
PCA(copy=True, iterated_power='auto', n_components=None, random_state=None,
    svd_solver='auto', tol=0.0, whiten=False)
```

管道中的评估器参数可以通过 <estimator>__<parameter> 格式来访问：

```
>>> pipe.set_params(clf__C=10)
Pipeline(memory=None,
        steps=[('reduce_dim', PCA(copy=True, iterated_power='auto',...)),
               ('clf', SVC(C=10, cache_size=200, class_weight=None,...))])
```

named_steps 的属性被映射到多个值：

```
>>> pipe.named_steps.reduce_dim is pipe.named_steps['reduce_dim']
True
```

单独的步骤可以用多个参数替换，除了最后步骤，其他步骤都可以设置为 None 来跳过：

```
>>> from sklearn.linear_model import LogisticRegression
>>> param_grid = dict(reduce_dim=[None, PCA(5), PCA(10)],
...                   clf=[SVC(), LogisticRegression()],
...                   clf__C=[0.1, 10, 100])
>>> grid_search = GridSearchCV(pipe, param_grid=param_grid)
```

⚠️ **注意** 对管道调用方法 fit 的效果跟依次对每个评估器调用方法 fit 一样，都是用 transform 输入并传递给下一个步骤。如果最后的评估器是一个分类器，管道可以当作分类器来用。如果最后一个评估器是转换器，管道也一样可以当作转换器。

请看下面的实例文件 Pipeline01.py，功能是使用 Pipeline 实现特征提取和评估。本实例使用的是拥有 20 个新闻组的数据集数，执行后将自动下载这个数据集，然后缓存并重新用于文档分类处理功能。

```
from sklearn.pipeline import Pipeline

print(__doc__)

#在标准输出上显示进度日志
logging.basicConfig(level=logging.INFO,
                    format='%(asctime)s %(levelname)s %(message)s')

# #################################################################
#从训练集中加载某些类别
categories = [
    'alt.atheism',
    'talk.religion.misc',
]
#设置为categories = None,以对所有类别进行分析
```

```python
print("Loading 20 newsgroups dataset for categories:")
print(categories)

data = fetch_20newsgroups(subset='train', categories=categories)
print("%d documents" % len(data.filenames))
print("%d categories" % len(data.target_names))
print()

# ##########################################################################
#定义一个结合文本特征提取器和简单分类器的管道
pipeline = Pipeline([
    ('vect', CountVectorizer()),
    ('tfidf', TfidfTransformer()),
    ('clf', SGDClassifier()),
])

#取消注释更多参数将提供更好的探索能力,但会以组合方式增加处理时间
parameters = {
    'vect__max_df': (0.5, 0.75, 1.0),
    # 'vect__max_features': (None, 5000, 10000, 50000),
    'vect__ngram_range': ((1, 1), (1, 2)),  # vect-ngram 参数
    # 'tfidf__use_idf': (True, False),
    # 'tfidf__norm': ('l1', 'l2'),
    'clf__max_iter': (20,),
    'clf__alpha': (0.00001, 0.000001),
    'clf__penalty': ('l2', 'elasticnet'),
    # 'clf__max_iter': (10, 50, 80),
}

if __name__ == "__main__":
    #多重处理要求fork发生在 __main__ protected块中
    #为特征提取和分类器寻找最佳参数
    grid_search = GridSearchCV(pipeline, parameters, n_jobs=-1, verbose=1)

    print("Performing grid search...")
    print("pipeline:", [name for name, _ in pipeline.steps])
    print("parameters:")
    pprint(parameters)
    t0 = time()
    grid_search.fit(data.data, data.target)
    print("done in %0.3fs" % (time() - t0))
    print()

    print("Best score: %0.3f" % grid_search.best_score_)
    print("Best parameters set:")
    best_parameters = grid_search.best_estimator_.get_params()
    for param_name in sorted(parameters.keys()):
        print("\t%s: %r" % (param_name, best_parameters[param_name]))
```

执行后会输出下面的结果，也可以将类别名称提供给数据集加载器。

```
Loading 20 newsgroups dataset for categories:
['alt.atheism', 'talk.religion.misc']
1427 documents
2 categories

Performing grid search...
pipeline: ['vect', 'tfidf', 'clf']
parameters:
{'clf__alpha': (1.0000000000000001e-05, 9.9999999999999995e-07),
 'clf__max_iter': (10, 50, 80),
 'clf__penalty': ('l2', 'elasticnet'),
 'tfidf__use_idf': (True, False),
 'vect__max_n': (1, 2),
 'vect__max_df': (0.5, 0.75, 1.0),
 'vect__max_features': (None, 5000, 10000, 50000)}
done in 1737.030s

Best score: 0.940
Best parameters set:
    clf__alpha: 9.9999999999999995e-07
    clf__max_iter: 50
    clf__penalty: 'elasticnet'
    tfidf__use_idf: True
    vect__max_n: 2
    vect__max_df: 0.75
    vect__max_features: 50000
```

6.1.2 FeatureUnion（特征联合）：特征层

在 Scikit-Learn 应用中，特征联合（FeatureUnion）合并了多个转换器对象形成一个新的转换器，该转换器合并了它们的输出。一个特征联合可以接收多个转换器对象。在适配期间，每个转换器都单独地和数据适配。对于转换数据，转换器可以并发使用，且输出的样本向量被连接成更大的向量。

特征联合的功能与 Pipeline 一样，也是为了实现便捷性和联合参数的估计和验证，可以结合 class:FeatureUnion 和 Pipeline 创造出复杂模型。

在 Scikit-Learn 应用中，通过一系列键值对 (key, value) 来构建特征联合，其中，key 是给转换器指定的名字（一个绝对的字符串，它只是一个代号），value 是一个评估器对象，例如：

```
>>> from sklearn.pipeline import FeatureUnion
>>> from sklearn.decomposition import PCA
>>> from sklearn.decomposition import KernelPCA
>>> estimators = [('linear_pca', PCA()), ('kernel_pca', KernelPCA())]
>>> combined = FeatureUnion(estimators)
>>> combined
FeatureUnion(n_jobs=1,
```

```
transformer_list=[('linear_pca', PCA(copy=True,...)),
                  ('kernel_pca', KernelPCA(alpha=1.0,...))],
transformer_weights=None)
```

跟管道一样，特征联合有一个精简版的构造器：unc:<cite>make_union</cite>，该构造器不需要显式给每个组件起名字。正如 Pipeline 那样，单独的步骤可以用 set_params 来替换，并设置为 None 来跳过：

```
>>> combined.set_params(kernel_pca=None)
...
FeatureUnion(n_jobs=1,
 transformer_list=[('linear_pca', PCA(copy=True,...)),
                   ('kernel_pca', None)],
 transformer_weights=None)
```

请看下面的实例文件 tiqu.py，演示了串联多个特征提取的方法。在现实例子中，有很多方法可以从数据集中提取特征，人们通常将几种方法结合起来以获得良好的性能。本实例展示了如何使用 FeatureUnionPCA 和单变量选择获得的特征然后进行组合的用法。

```python
iris = load_iris()
X, y = iris.data, iris.target
#PCA
pca = PCA(n_components=2)
selection = SelectKBest(k=1)

#利用主成分分析和单变量选择建立估计量
combined_features = FeatureUnion([("pca", pca), ("univ_select", selection)])
#使用组合特征转换数据集
X_features = combined_features.fit(X, y).transform(X)
print("Combined space has", X_features.shape[1], "features")

svm = SVC(kernel="linear")
#对 k、n_components 和 C 执行网格搜索

pipeline = Pipeline([("features", combined_features), ("svm", svm)])

param_grid = dict(features__pca__n_components=[1, 2, 3],
                  features__univ_select__k=[1, 2],
                  svm__C=[0.1, 1, 10])

grid_search = GridSearchCV(pipeline, param_grid=param_grid, verbose=10)
grid_search.fit(X, y)
print(grid_search.best_estimator_)
```

在上述代码中，使用此转换器组合功能的好处是允许在整个过程中进行交叉验证和网格搜索。执行后会输出：

```
Combined space has 3 features
Fitting 5 folds for each of 18 candidates, totalling 90 fits
[CV 1/5; 1/18] START features__pca__n_components=1, features__univ_select__k=1, svm__C=0.1
```

```
    [CV 1/5; 1/18] END features__pca__n_components=1, features__univ_select__k=1,
svm__C=0.1;, score=0.933 total time=   0.0s
    [CV 2/5; 1/18] START features__pca__n_components=1, features__univ_select__k=1, svm__C=0.1
    ……
    省略其他输出结果
    ……
    [CV 3/5; 18/18] START features__pca__n_components=3, features__univ_select__k=2, svm__C=10
    [CV 3/5; 18/18] END features__pca__n_components=3, features__univ_select__k=2,
svm__C=10;, score=0.900 total time=   0.0s
    [CV 4/5; 18/18] START features__pca__n_components=3, features__univ_select__k=2, svm__C=10
    [CV 4/5; 18/18] END features__pca__n_components=3, features__univ_select__k=2,
svm__C=10;, score=0.967 total time=   0.0s
    [CV 5/5; 18/18] START features__pca__n_components=3, features__univ_select__k=2, svm__C=10
    [CV 5/5; 18/18] END features__pca__n_components=3, features__univ_select__k=2,
svm__C=10;, score=1.000 total time=   0.0s
    Pipeline(steps=[('features',
                    FeatureUnion(transformer_list=[('pca', PCA(n_components=3)),
                                                  ('univ_select',
                                                    SelectKBest(k=1))])),
                    ('svm', SVC(C=10, kernel='linear'))])
```

> **注意** 在本实例中使用的组合对该数据集没有特别的帮助,仅用于说明 FeatureUnion 的用法。

6.2 特征提取

在 Scikit-Learn 应用中,模块 sklearn.feature_extraction 能够提取符合机器学习算法支持的特征,比如文本和图片。特征提取与特征选择有很大的不同:前者包括将任意数据(如文本或图像)转换为可用于机器学习的数值特征,后者是将这些特征应用到机器学习中。

扫码观看本节视频讲解

6.2.1 从字典类型加载特征

在 Scikit-Learn 应用中,类 DictVectorizer 能够将标准的 Python 字典(dict)对象列表的要素数组转换为 Scikit-Learn 估计器使用的表示形式"NumPy/SciPy"。虽然 Python 的处理速度不是特别快,但是 Python 的 dict 优点是使用方便,并且除了值之外还存储特征名称。

类 DictVectorizer 实现了"one-of-K"或"one-hot"编码,用于分类(也称为标称、离散)特征。分类功能是"属性值"对,其中该值被限制为没有排序的离散可能性列表(例如主题标识符、对象类型、标签、名称……)。

例如下面的例子,"城市"是一个分类属性,而"温度"是传统的数字特征:

```
>>> measurements = [
```

```
...         {'city': 'Dubai', 'temperature': 33.},
...         {'city': 'London', 'temperature': 12.},
...         {'city': 'San Francisco', 'temperature': 18.},
... ]
>>> from sklearn.feature_extraction import DictVectorizer
>>> vec = DictVectorizer()

>>> vec.fit_transform(measurements).toarray()
array([[ 1.,  0.,  0., 33.],
       [ 0.,  1.,  0., 12.],
       [ 0.,  0.,  1., 18.]])

>>> vec.get_feature_names()
['city=Dubai', 'city=London', 'city=San Francisco', 'temperature']
```

类 DictVectorizer 也是对自然语言处理模型中训练序列分类器的有用的变换工具，通常通过提取围绕感兴趣的特定词的特征窗口来工作。

6.2.2 特征哈希

在 Scikit-Learn 中通过类 FeatureHasher 实现特征哈希功能。特征哈希是一种高速、低内存消耗的向量化方法，它使用了特征散列（feature hashing）技术 [或被称为散列法（hashing trick）技术]，代替在构建训练中遇到的特征的哈希表，如向量化所做的那样，FeatureHasher 将哈希函数应用于特征，以便直接在样本矩阵中确定它们的列索引。这是以牺牲可检测性为代价，目的是提高速度和减少内存的使用。

由于散列函数可能导致（不相关）特征之间的冲突，因此使用带符号散列函数，并且散列值的符号确实是存储在特征输出矩阵中。这样，冲突可能会抵消错误，并且任何输出要素的值的预期平均值为零。在默认情况下，此机制将设置使用 alternate_sign=True，这对于小型哈希表（n_features < 10000）特别有用。对于大的哈希表，可以禁用它，以便将输出传递给估计器，如 sklearn.naive_bayes.MultinomialNB 或 sklearn.feature_selection.chi2 特征选择器，这些特征选择器可以使用非负输入。

类 FeatureHasher 接受映射（如 Python 的 dict 及其在 collections 模块中的变体），使用键值对（feature，value）或字符串，具体取决于构造函数参数 input_type。 映射被视为（feature，value）对的列表，而单个字符串的隐含值为 1，因此 ['feat1','feat2','feat3'] 被解释为 [('feat1', 1), ('feat2', 1), ('feat3', 1)]。如果在样本中多次出现单个特征，相关值将求和 [所以 ('feat', 2) 和 ('feat', 3.5) 变为 ('feat', 5.5)]。FeatureHasher 的输出始终是 CSR 格式的 scipy.sparse 矩阵。

可以在文档分类中使用特征散列，但与 text.CountVectorizer 不同，FeatureHasher 不执行除 Unicode 或 UTF-8 编码之外的任何其他预处理。例如，有一个词级别的自然语言处理任务，需要从 (token, part_of_speech) 键值对中提取特征，此时可以使用 Python 生成器函数来提取功能：

```
def token_features(token, part_of_speech):
    if token.isdigit():
        yield "numeric"
```

```
        else:
            yield "token={}".format(token.lower())
            yield "token,pos={},{}".format(token, part_of_speech)
        if token[0].isupper():
            yield "uppercase_initial"
        if token.isupper():
            yield "all_uppercase"
        yield "pos={}".format(part_of_speech)
```

然后，raw_X 为了可以传入 FeatureHasher.transform，可以通过如下方式进行构造：

```
raw_X = (token_features(tok, pos_tagger(tok)) for tok in corpus)
```

接着传入一个 hasher：

```
hasher = FeatureHasher(input_type='string')
X = hasher.transform(raw_X)
```

这样会得到一个 scipy.sparse 类型的矩阵 X。

6.2.3 提取文本特征

文本分析是机器学习算法的主要应用领域之一，然而原始数据和符号文字序列不能直接被传递给算法，因为它们大多数要求具有固定长度的数字矩阵特征向量，而不是具有可变长度的原始文本文档。

为了解决这个问题，Scikit-Learn 提供了从文本内容中提取数字特征的最常见方法。

- 令牌化（tokenizing）：将每个可能的词令牌设置成字符串并赋予整数型的id，例如，通过使用空格和标点符号作为令牌分隔符。
- 统计（counting）：每个词令牌在文档中出现的次数。
- 标准化（normalizing）：在大多数的"文档/样本"中，可以减少重要的次令牌的出现次数的权重。

在该方案中，对特征和样本定义如下。

- 每个"单独的令牌发生频率"（归一化或不归零）被视为一个"特征"。
- 给定"文档"中所有的令牌频率向量被看作一个多元sample"样本"。

因此，文本的集合可被表示为矩阵形式，每行对应一条文本，每列对应每个文本中出现的词令牌（如单个词）。

我们称"向量化"是将文本文档集合转换为数字集合特征向量的普通方法。这种特殊思想（令牌化、计数和归一化）被称为"Bag of Words"或"Bag of n-grams"（词袋）模型。

在 Scikit-Learn 应用中，类 CountVectorizer 在单个类中实现了令牌化和出现频数统计：

```
>>> from sklearn.feature_extraction.text import CountVectorizer
```

这个模型中有很多参数，但参数的默认初始值是相当合理的：

```
>>> vectorizer = CountVectorizer()
>>> vectorizer
```

```
CountVectorizer(analyzer=...'word', binary=False, decode_error=...'strict',
        dtype=<... 'numpy.int64'>, encoding=...'utf-8', input=...'content',
        lowercase=True, max_df=1.0, max_features=None, min_df=1,
        ngram_range=(1, 1), preprocessor=None, stop_words=None,
        strip_accents=None, token_pattern=...'(?u)\\b\\w\\w+\\b',
        tokenizer=None, vocabulary=None)
```

接下来使用它进行简单的文本全集令牌化处理，并统计词频：

```
>>> corpus = [
...     'This is the first document.',
...     'This is the second second document.',
...     'And the third one.',
...     'Is this the first document?',
... ]
>>> X = vectorizer.fit_transform(corpus)
>>> X
<4x9 sparse matrix of type '<... 'numpy.int64'>'
    with 19 stored elements in Compressed Sparse ... format>
```

开始令牌化字符串，提取至少两个字母的词默认配置：

```
>>> analyze = vectorizer.build_analyzer()
>>> analyze("This is a text document to analyze.") == (
...     ['this', 'is', 'text', 'document', 'to', 'analyze'])
True
```

函数 analyzer 在拟合期间发现的每个项都被分配一个与所得矩阵中的列对应的唯一整数索引。下面是对列的检索过程：

```
>>> vectorizer.get_feature_names() == (
...     ['and', 'document', 'first', 'is', 'one',
...      'second', 'the', 'third', 'this'])
True

>>> X.toarray()
array([[0, 1, 1, 1, 0, 0, 1, 0, 1],
       [0, 1, 0, 1, 0, 2, 1, 0, 1],
       [1, 0, 0, 0, 1, 0, 1, 1, 0],
       [0, 1, 1, 1, 0, 0, 1, 0, 1]]...)
```

从列标到特征名的反转映射存储在向量化类 vectorizer 的属性 vocabulary_ 中：

```
>>> vectorizer.vocabulary_.get('document')
1
```

在将来的调用转换方法中，训练语料库中未出现的词将被完全忽略：

```
>>> vectorizer.transform(['Something completely new.']).toarray()
...
```

```
array([[0, 0, 0, 0, 0, 0, 0, 0, 0]]...)
```

请注意,在上一个语料库中,第一个和最后一个文档具有完全相同的单词,因此被编码成相同的向量。特别是丢失的最后一个字符是疑问形式的信息。为了防止词组顺序颠倒,除了提取一元模型 1-grams(个别词)之外,还可以提取二元模型 2-grams 的单词:

```
>>> bigram_vectorizer = CountVectorizer(ngram_range=(1, 2),
...                                     token_pattern=r'\b\w+\b', min_df=1)
>>> analyze = bigram_vectorizer.build_analyzer()
>>> analyze('Bi-grams are cool!') == (
...     ['bi', 'grams', 'are', 'cool', 'bi grams', 'grams are', 'are cool'])
True
```

矢量化提取的词因此变得很大,同时可以在定位模式时消除歧义:

```
>>> X_2 = bigram_vectorizer.fit_transform(corpus).toarray()
>>> X_2
...
array([[0, 0, 1, 1, 1, 1, 1, 0, 0, 0, 0, 0, 1, 1, 0, 0, 0, 0, 1, 1, 0],
       [0, 0, 1, 0, 0, 1, 1, 0, 0, 2, 1, 1, 1, 0, 1, 0, 0, 0, 1, 1, 0],
       [1, 1, 0, 0, 0, 0, 0, 0, 1, 0, 0, 0, 1, 0, 0, 1, 1, 1, 0, 0, 0],
       [0, 0, 1, 1, 1, 1, 0, 1, 0, 0, 0, 0, 1, 1, 0, 0, 0, 0, 1, 0, 1]]...)
```

特别是"is this"的疑问形式只出现在最后一个文档中:

```
>>> feature_index = bigram_vectorizer.vocabulary_.get('is this')
>>> X_2[:, feature_index]
array([0, 0, 0, 1]...)
```

6.2.4 提取图像特征

在 Scikit-Learn 应用中,函数 extract_patches_2d(image, patch_size, max_patches=None, random_state=None) 的作用是按照给定尺寸的要求,随机提取原始图像 patch 大小尺寸图片,然后将其返回。patch_size 就是要提取的尺寸。max_patches 是 0 和 1 之间的数,表示提取 pathes 的比例,如果是 1 就表示提取整个 patch。例如,下面是使用 3 个彩色通道(例如 RGB 格式)生成一个 4×4 像素的图像的代码:

```
>>> import numpy as np
>>> from sklearn.feature_extraction import image

>>> one_image = np.arange(4 * 4 * 3).reshape((4, 4, 3))
>>> one_image[:, :, 0]  # RGB图片的R通道
array([[ 0,  3,  6,  9],
       [12, 15, 18, 21],
       [24, 27, 30, 33],
       [36, 39, 42, 45]])

>>> patches = image.extract_patches_2d(one_image, (2, 2), max_patches=2,
```

```
...        random_state=0)
>>> patches.shape
(2, 2, 2, 3)
>>> patches[:, :, :, 0]
array([[[ 0,  3],
        [12, 15]],

       [[15, 18],
        [27, 30]]])
>>> patches = image.extract_patches_2d(one_image, (2, 2))
>>> patches.shape
(9, 2, 2, 3)
>>> patches[4, :, :, 0]
array([[15, 18],
       [27, 30]])
```

现在尝试通过在重叠区域进行平均来从补丁重建原始图像：

```
>>> reconstructed = image.reconstruct_from_patches_2d(patches, (4, 4, 3))
>>> np.testing.assert_array_equal(one_image, reconstructed)
```

在 Scikit-Learn 应用中，还可以使用特征或样本之间的连接信息。例如，通过 Ward 聚类（分层聚类）可以将只有图像的相邻像素形成连续的斑块。函数 img_to_graph 能够从 2D 或 3D 图像返回一个对应的矩阵。类似地，函数 grid_to_graph 能够为指定形状的图像构建连接矩阵。这些矩阵可用于在使用连接信息的估计器中强加连接，如 Ward 聚类（分层聚类），而且还要构建预计算的内核或相似矩阵。

请看下面的实例文件 ju.py，功能是在硬币图像中构建结构化 Ward 层次聚类。

```python
import skimage
from skimage.data import coins
from skimage.transform import rescale

from sklearn.feature_extraction.image import grid_to_graph
from sklearn.cluster import AgglomerativeClustering
from sklearn.utils.fixes import parse_version

if parse_version(skimage.__version__) >= parse_version('0.14'):
    rescale_params = {'anti_aliasing': False, 'multichannel': False}
else:
    rescale_params = {}

# #########################################################################
#生成数据
orig_coins = coins()

#将其调整为原始大小的20%,以加快处理速度在缩小比例之前应用高斯滤波器进行平滑,从而减少混叠伪影
smoothened_coins = gaussian_filter(orig_coins, sigma=2)
rescaled_coins = rescale(smoothened_coins, 0.2, mode="reflect",
                         **rescale_params)
```

```python
X = np.reshape(rescaled_coins, (-1, 1))

# ##############################################################################
#定义数据的结构,用像素连接到它们的邻居
connectivity = grid_to_graph(*rescaled_coins.shape)

# ##############################################################################
#计算聚类
print("Compute structured hierarchical clustering...")
st = time.time()
n_clusters = 27
#区域数
ward = AgglomerativeClustering(n_clusters=n_clusters, linkage='ward',
                               connectivity=connectivity)
ward.fit(X)
label = np.reshape(ward.labels_, rescaled_coins.shape)
print("Elapsed time: ", time.time() - st)
print("Number of pixels: ", label.size)
print("Number of clusters: ", np.unique(label).size)

# ##############################################################################
#将结果打印到图像上
plt.figure(figsize=(5, 5))
plt.imshow(rescaled_coins, cmap=plt.cm.gray)
for l in range(n_clusters):
    plt.contour(label == l,
                colors=[plt.cm.nipy_spectral(l / float(n_clusters)), ])
plt.xticks(())
plt.yticks(())
plt.show()
```

ju.py 的执行效果如图 6-1 所示。

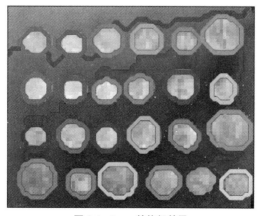

图 6-1 ju.py 的执行效果

6.3 预处理数据

在 Scikit-Learn 应用中,sklearn.preprocessing 包提供了几个常见的实用功能和变换器类型,用来将原始特征向量更改为更适合机器学习模型的形式。一般来说,机器学习算法受益于数据集的标准化。如果数据集中存在一些离群值,那么稳定的缩放或转换更合适。

扫码观看本节视频讲解

6.3.1 标准化处理

在 Scikit-Learn 应用中,数据集的标准化处理是常见的要求。如果个别特征或多或少看起来不是很像标准正态分布(具有零均值和单位方差),那么它们的表现力可能会较差。在实际应用中,我们经常忽略特征的分布形状,直接经过去均值来对某个特征进行中心化,再通过除以非常量特征(non-constant features)的标准差进行缩放。

例如,在机器学习算法的目标函数(例如,SVM 的 RBF 内核或线性模型的 L1 和 L2 正则化)中,许多学习算法中的目标函数的基础都是假设所有的特征都是零均值并且具有同一阶数上的方差。如果某个特征的方差比其他特征大几个数量级,那么它就会在学习算法中占据主导位置,导致学习器并不能像我们所期望的那样,从其他特征中进行学习。例如,函数 scale() 为数组形状的数据集的标准化提供了一个快捷实现:

```
>>> from sklearn import preprocessing
>>> import numpy as np
>>> X_train = np.array([[ 1., -1.,  2.],
...                     [ 2.,  0.,  0.],
...                     [ 0.,  1., -1.]])
>>> X_scaled = preprocessing.scale(X_train)

>>> X_scaled
array([[ 0.  ..., -1.22...,  1.33...],
       [ 1.22...,  0.  ..., -0.26...],
       [-1.22...,  1.22..., -1.06...]])
```

经过缩放后的数据具有零均值以及标准方差:

```
>>> X_scaled.mean(axis=0)
array([ 0., 0., 0.])

>>> X_scaled.std(axis=0)
array([ 1., 1., 1.])
```

在 Scikit-Learn 预处理模块中还提供了一个实用类 StandardScaler,它实现了用转化器的 API 来计算训练集上的平均值和标准偏差,以便以后能够在测试集上重新应用相同的变换。因此,这个类适用于 sklearn.pipeline.Pipeline 的早期步骤:

```
>>> scaler = preprocessing.StandardScaler().fit(X_train)
>>> scaler
StandardScaler(copy=True, with_mean=True, with_std=True)

>>> scaler.mean_
array([ 1. ...,  0. ...,  0.33...])

>>> scaler.scale_
array([ 0.81...,  0.81...,  1.24...])

>>> scaler.transform(X_train)
array([[ 0.  ..., -1.22...,  1.33...],
       [ 1.22...,  0.  ..., -0.26...],
       [-1.22...,  1.22..., -1.06...]])
```

缩放类对象可以在新的数据上实现和训练集相同的缩放操作：

```
>>> X_test = [[-1., 1., 0.]]
>>> scaler.transform(X_test)
array([[-2.44...,  1.22..., -0.26...]])
```

也可以通过在构造函数 class:StandardScaler 中传入参数 with_mean=False 或者 with_std=False 来取消中心化或缩放操作。

6.3.2 非线性转换

类似于缩放，类 QuantileTransformer 将每个特征缩放在同样的范围或分布情况下。但是，通过执行一个秩转换能够使异常的分布平滑化，并且能够比缩放更少地受到离群值的影响。但是它的确使特征间及特征内的关联和距离失真了。

类 QuantileTransformer 以及函数 quantile_transform 提供了一个基于分位数函数的无参数转换，将数据映射到了 0~1 的均匀分布上：

```
>>> from sklearn.datasets import load_iris
>>> from sklearn.model_selection import train_test_split
>>> iris = load_iris()
>>> X, y = iris.data, iris.target
>>> X_train, X_test, y_train, y_test = train_test_split(X, y, random_state=0)
>>> quantile_transformer = preprocessing.QuantileTransformer(random_state=0)
>>> X_train_trans = quantile_transformer.fit_transform(X_train)
>>> X_test_trans = quantile_transformer.transform(X_test)
>>> np.percentile(X_train[:, 0], [0, 25, 50, 75, 100])
array([4.3, 5.1, 5.8, 6.5, 7.9])
```

这个特征是萼片的长度，单位为厘米。一旦应用分位数转换，这些元素就接近于之前定义的百分位数：

```
>>> np.percentile(X_train_trans[:, 0], [0, 25, 50, 75, 100])
...
```

```
array([ 0.00..., 0.24..., 0.49..., 0.73..., 0.99... ])
```

这可以在具有类似形式的独立测试集上进行确认：

```
>>> np.percentile(X_test[:, 0], [0, 25, 50, 75, 100])
...
array([ 4.4, 5.125, 5.75, 6.175, 7.3])
>>> np.percentile(X_test_trans[:, 0], [0, 25, 50, 75, 100])
...
array([ 0.01..., 0.25..., 0.46..., 0.60..., 0.94...])
```

也可以通过设置 output_distribution='normal' 将转换后的数据映射到正态分布：

```
>>> quantile_transformer = preprocessing.QuantileTransformer(
...     output_distribution='normal', random_state=0)
>>> X_trans = quantile_transformer.fit_transform(X)
>>> quantile_transformer.quantiles_
array([[ 4.3...,   2...,    1...,    0.1...],
       [ 4.31...,  2.02..., 1.01..., 0.1...],
       [ 4.32...,  2.05..., 1.02..., 0.1...],
       ...,
       [ 7.84...,  4.34..., 6.84..., 2.5...],
       [ 7.87...,  4.37..., 6.87..., 2.5...],
       [ 7.9...,   4.4...,  6.9...,  2.5...]])
```

将输入的中值称为输出的平均值，并且以 0 为中心。正常输出被剪切，使得输入的最小值和最大值分别对应于 1e-7 和 1-1e-7 分位数：在变换下不会变得无限大。

请看下面的实例文件 lisan.py，功能是使用 KBinsDiscretizer 实现离散化连续特征处理。本实例比较了线性回归（线性模型）和决策树（基于树的模型）在实值特征离散化和未离散化的情况下的预测结果。

```
#构建数据集
rnd = np.random.RandomState(42)
X = rnd.uniform(-3, 3, size=100)
y = np.sin(X) + rnd.normal(size=len(X)) / 3
X = X.reshape(-1, 1)

#用KBinsDiscretizer变换数据集
enc = KBinsDiscretizer(n_bins=10, encode='onehot')
X_binned = enc.fit_transform(X)

#用原始数据集预测
fig, (ax1, ax2) = plt.subplots(ncols=2, sharey=True, figsize=(10, 4))
line = np.linspace(-3, 3, 1000, endpoint=False).reshape(-1, 1)
reg = LinearRegression().fit(X, y)
ax1.plot(line, reg.predict(line), linewidth=2, color='green',
         label="linear regression")
reg = DecisionTreeRegressor(min_samples_split=3, random_state=0).fit(X, y)
ax1.plot(line, reg.predict(line), linewidth=2, color='red',
         label="decision tree")
ax1.plot(X[:, 0], y, 'o', c='k')
```

```
ax1.legend(loc="best")
ax1.set_ylabel("Regression output")
ax1.set_xlabel("Input feature")
ax1.set_title("Result before discretization")

#利用转换数据集进行预测
line_binned = enc.transform(line)
reg = LinearRegression().fit(X_binned, y)
ax2.plot(line, reg.predict(line_binned), linewidth=2, color='green',
         linestyle='-', label='linear regression')
reg = DecisionTreeRegressor(min_samples_split=3,
                            random_state=0).fit(X_binned, y)
ax2.plot(line, reg.predict(line_binned), linewidth=2, color='red',
         linestyle=':', label='decision tree')
ax2.plot(X[:, 0], y, 'o', c='k')
ax2.vlines(enc.bin_edges_[0], *plt.gca().get_ylim(), linewidth=1, alpha=.2)
ax2.legend(loc="best")
ax2.set_xlabel("Input feature")
ax2.set_title("Result after discretization")

plt.tight_layout()
plt.show()
```

lisan.py 执行效果如图 6-2 所示，如离散化前的结果所示，线性模型构建速度快且解释相对简单，但只能建模线性关系，而决策树可以构建更复杂的数据模型。使线性模型在连续数据上更强大的一种方法是使用离散化（也称为分箱）。在本实例中，对特征进行离散化处理，并对转换后的数据进行热编码处理。请注意，如果 bin 的宽度不合理，则过度拟合的风险可能会大大增加，因此通常应在交叉验证下调整离散化器参数。

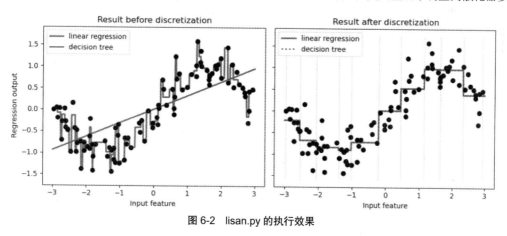

图 6-2　lisan.py 的执行效果

在离散化后，线性回归和决策树做出完全相同的预测。由于每个 bin 内的特征是恒定的，因此任何模型都必须为 bin 内的所有点预测相同的值。与离散化前的结果相比，线性模型变得更加灵活，而决策树变得更加不灵活。请注意，分箱特征通常对基于树的模型没有任何好处，因为这些模型可以学习在任何地方拆分数据。

6.4 无监督降维

在 Scikit-Learn 应用中，如果特征数量很多，在监督之前可以通过无监督来减少特征。很多的无监督学习方法实现了一个名为 transform 的方法，可以用此方法来降低维度。

扫码观看本节视频讲解

6.4.1 PCA：主成分分析

在 Scikit-Learn 应用中，decomposition.PCA 能够捕捉原始特征的差异特征的组合，其原型如下：

```
class sklearn.decomposition.PCA(n_components=None, *, copy=True, whiten=False, svd_solver='auto', tol=0.0, iterated_power='auto', random_state=None)
```

其中的参数说明如下。

（1）n_components：这个参数可以帮我们指定希望 PCA 降维后的特征维度数目。最常用的做法是直接指定降维到的维度数目，此时 n_components 是一个大于等于 1 的整数。当然，也可以指定主成分的方差和所占的最小比例阈值，让 PCA 类自己去根据样本特征方差来决定降维到的维度数，此时 n_components 是一个 (0,1] 之间的数。当然，我们还可以将参数设置为"mle"，此时 PCA 类会用 MLE 算法根据特征的方差分布情况自己去选择一定数量的主成分特征来降维。我们也可以用默认值，即不输入 n_components，此时 n_components=min（样本数，特征数）。

（2）copy：bool 类型，取值是 True 或者 False，缺省时默认为 True，表示是否在运行算法时将原始训练数据复制一份。若为 True，运行 PCA 算法后，原始训练数据的值不会有任何改变，因为是在原始数据的副本上进行运算；若为 False，则运行 PCA 算法后，原始训练数据的值会改变，因为是在原始数据上进行降维计算。

（3）whiten：判断是否进行白化。所谓白化，就是对降维后的数据的每个特征进行归一化，让方差都为 1。对于 PCA 降维本身来说，一般不需要白化。如果 PCA 降维后有后续的数据处理动作，可以考虑白化。默认值是 False，即不进行白化。

（4）svd_solver：指定奇异值分解 SVD 的方法，由于特征分解是奇异值分解 SVD 的一个特例，一般的 PCA 库都是基于 SVD 实现的。有 4 个可以选择的值：{'auto', 'full', 'arpack', 'randomized'}。randomized 一般适用于数据量大、数据维度多同时主成分数目比例又较低的 PCA 降维，它使用了一些加快 SVD 的随机算法。full 则是传统意义上的 SVD，使用了 scipy 库的实现方法。arpack 和 randomized 的适用场景类似，区别是 randomized 使用的是 Scikit-Learn 自己的 SVD 实现，而 arpack 直接使用了 scipy 库的 sparse SVD 实现。默认是 auto，即 PCA 类会自己在前面讲到的三种算法里面去权衡，选择一个合适的 SVD 算法来降维。一般来说，使用默认值就够了。

（5）tol：svd_solver =='arpack' 计算的奇异值的公差，float $\geq$ 0，可选（默认值是 0）。

（6）iterated_power：取值为大于等于 0 的整数或 'auto'，默认为 'auto'，当 svd_solver 的值为 randomized 时计算出幂方法的迭代次数。

在 decomposition.PCA 中，各个属性的具体说明如下。

- components_：特征空间中的主轴，表示数据中最大方差的方向，组件按 explained_variance_ 排序。

- explained_variance_：它代表降维后的各主成分的方差值。方差值越大，则说明越是重要的主成分。
- explained_variance_ratio_：它代表降维后的各主成分的方差值占总方差值的比例，这个比例越大，则越是重要的主成分。
- singular_values_：每个特征的奇异值，奇异值等于n_components 低维空间中变量的2范数。
- mean_：每个特征的均值。
- n_components_：即上面输入的参数值。
- n_features_：训练数据中的特征数量。
- n_samples_：训练数据中的样本数。
- noise_variance_：等于X协方差矩阵的(min(n_features,n_samples)-n_components)个最小特征值的平均值。

在 decomposition.PCA 中，各个方法的具体说明如下。

- fit(X [,y])：用X拟合模型。
- fit_transform(X [,y])：使用X拟合模型，并在X上应用降维。
- get_covariance()：用生成模型计算数据协方差。
- get_params([deep])：获取此估计量的参数。
- get_precision()：用生成模型计算数据精度矩阵。
- inverse_transform(X)：将数据转换回其原始空间。
- score(X [,y])：返回所有样本的平均对数似然率。
- score_samples(X)：返回每个样本的对数似然。
- set_params(**params)：设置此估算器的参数。
- transform(X)：对X应用降维。

请看下面的实例文件，功能是使用特征脸和 SVM 实现人脸识别。

```
#在标准输出上显示进度日志
logging.basicConfig(level=logging.INFO, format='%(asctime)s %(message)s')

# ###########################################################################
#下载数据(如果尚未在磁盘上)并将其作为numpy数组加载

lfw_people = fetch_lfw_people(min_faces_per_person=70, resize=0.4)

#缺省图像阵列以找到形状(用于绘图)
n_samples, h, w = lfw_people.images.shape

#对于机器学习,我们直接使用两个数据(因为此模型忽略了相对像素位置信息)
X = lfw_people.data
n_features = X.shape[1]

#要预测的标签是人的id
y = lfw_people.target
target_names = lfw_people.target_names
n_classes = target_names.shape[0]
```

```python
print("Total dataset size:")
print("n_samples: %d" % n_samples)
print("n_features: %d" % n_features)
print("n_classes: %d" % n_classes)

# ###############################################################################
#使用层分为训练集和测试集
X_train, X_test, y_train, y_test = train_test_split(
    X, y, test_size=0.25, random_state=42)

# ###############################################################################
#在人脸数据集(视为未标记数据集)上计算PCA(特征脸):无监督特征提取/降维
n_components = 150

print("Extracting the top %d eigenfaces from %d faces"
      % (n_components, X_train.shape[0]))
t0 = time()
pca = PCA(n_components=n_components, svd_solver='randomized',
          whiten=True).fit(X_train)
print("done in %0.3fs" % (time() - t0))

eigenfaces = pca.components_.reshape((n_components, h, w))

print("Projecting the input data on the eigenfaces orthonormal basis")
t0 = time()
X_train_pca = pca.transform(X_train)
X_test_pca = pca.transform(X_test)
print("done in %0.3fs" % (time() - t0))

# ###############################################################################
#训练SVM分类模型

print("Fitting the classifier to the training set")
t0 = time()
param_grid = {'C': [1e3, 5e3, 1e4, 5e4, 1e5],
              'gamma': [0.0001, 0.0005, 0.001, 0.005, 0.01, 0.1], }
clf = GridSearchCV(
    SVC(kernel='rbf', class_weight='balanced'), param_grid
)
clf = clf.fit(X_train_pca, y_train)
print("done in %0.3fs" % (time() - t0))
print("Best estimator found by grid search:")
print(clf.best_estimator_)

# ###############################################################################
#测试集模型质量的定量评价

print("Predicting people's names on the test set")
t0 = time()
```

```python
y_pred = clf.predict(X_test_pca)
print("done in %0.3fs" % (time() - t0))

print(classification_report(y_test, y_pred, target_names=target_names))
print(confusion_matrix(y_test, y_pred, labels=range(n_classes)))

# #############################################################################
#利用matplotlib对预测进行定性评估

def plot_gallery(images, titles, h, w, n_row=3, n_col=4):
    """Helper function to plot a gallery of portraits"""
    plt.figure(figsize=(1.8 * n_col, 2.4 * n_row))
    plt.subplots_adjust(bottom=0, left=.01, right=.99, top=.90, hspace=.35)
    for i in range(n_row * n_col):
        plt.subplot(n_row, n_col, i + 1)
        plt.imshow(images[i].reshape((h, w)), cmap=plt.cm.gray)
        plt.title(titles[i], size=12)
        plt.xticks(())
        plt.yticks(())

#在测试集的一部分绘制预测结果

def title(y_pred, y_test, target_names, i):
    pred_name = target_names[y_pred[i]].rsplit(' ', 1)[-1]
    true_name = target_names[y_test[i]].rsplit(' ', 1)[-1]
    return 'predicted: %s\ntrue:      %s' % (pred_name, true_name)

prediction_titles = [title(y_pred, y_test, target_names, i)
                     for i in range(y_pred.shape[0])]

plot_gallery(X_test, prediction_titles, h, w)

#画出最有意义的面孔
eigenface_titles = ["eigenface %d" % i for i in range(eigenfaces.shape[0])]
plot_gallery(eigenfaces, eigenface_titles, h, w)

plt.show()
```

执行后会输出:

```
Total dataset size:
n_samples: 1288
n_features: 1850
n_classes: 7
Extracting the top 150 eigenfaces from 966 faces
done in 0.128s
Projecting the input data on the eigenfaces orthonormal basis
done in 0.015s
Fitting the classifier to the training set
```

```
done in 24.564s
Best estimator found by grid search:
SVC(C=1000.0, class_weight='balanced', gamma=0.005)
Predicting people's names on the test set
done in 0.074s
                   precision    recall  f1-score   support

     Ariel Sharon       0.75      0.46      0.57        13
     Colin Powell       0.81      0.87      0.84        60
  Donald Rumsfeld       0.86      0.67      0.75        27
    George W Bush       0.85      0.98      0.91       146
Gerhard Schroeder       0.95      0.80      0.87        25
      Hugo Chavez       1.00      0.60      0.75        15
       Tony Blair       0.97      0.81      0.88        36

         accuracy                           0.86       322
        macro avg       0.88      0.74      0.80       322
     weighted avg       0.87      0.86      0.85       322

[[  6   2   0   5   0   0   0]
 [  1  52   1   6   0   0   0]
 [  1   2  18   6   0   0   0]
 [  0   3   0 143   0   0   0]
 [  0   1   0   3  20   0   1]
 [  0   3   0   2   1   9   0]
 [  0   1   2   4   0   0  29]]
```

6.4.2 随机投影

在 Scikit-Learn 应用中，模块 random_projection 提供了几种用于通过随机投影减少数据的工具。sklearn.random_projection 模块实现了一种简单和计算高效的方法，通过交易控制量的精度（作为附加方差），以缩短数据的维数，从而缩短处理时间和缩小模型。该模块实现两种类型的非结构化随机矩阵：高斯随机矩阵和稀疏随机矩阵。

我们可以控制随机映射矩阵的维度和分布，这样可以保证在数据集中任意两个样本的距离。随机映射也是一种合适的基于距离的近似精确的方法。

请看下面的实例文件 tou.py，功能是实现用于嵌入随机投影的 Johnson-Lindenstrauss。Johnson-Lindenstrauss 定理缩写为 J-L 定理，其主要是说"一个 dd 维空间中的 nn 个点可以近似等距地嵌入一个 $k \approx O(\log n) k \approx O(\log n)$ 维的空间"，所谓近似等距的意思，简单地理解就是保持任意两个点之间的相对远近关系，一个不是十分确切的说法是拓扑同构。该定理是于 1984 年被发现的，在压缩感知、流行学习和降维上被应用。

实例文件 tou.py 的具体实现流程如下。

（1）随机投影引入的失真 p 由以下事实断言，p 即定义具有良好概率的 eps 嵌入，定义如下：

$$(1-\text{eps})\|u-v\|^2 < \|p(u)-p(v)\|^2 < (1+\text{eps})\|u-v\|^2$$

其中，u 和 v 是从形状为 (n_samples, n_features) 的数据集中提取的任何行，而 p 是形状为 (n_components, n_features) 的随机高斯 $N(0, 1)$ 矩阵（或稀疏的 Achlioptas 矩阵）的投影。

保证 eps 嵌入的最小组件数由如下式子设置：

$$\text{n\_components} \geq 4\log(\text{n\_samples})/\left(\text{eps}^2/2 - \text{eps}^3/3\right)$$

绘制变化范围，随着样本 n_samples 数量的 n_components 增加，最小维数以对数方式增加，以保证 eps 嵌入。代码如下：

```
#变化范围
eps_range = np.linspace(0.1, 0.99, 5)
colors = plt.cm.Blues(np.linspace(0.3, 1.0, len(eps_range)))

#要嵌入的样本数(观察值)范围
n_samples_range = np.logspace(1, 9, 9)

plt.figure()
for eps, color in zip(eps_range, colors):
    min_n_components = johnson_lindenstrauss_min_dim(n_samples_range, eps=eps)
    plt.loglog(n_samples_range, min_n_components, color=color)

plt.legend(["eps = %0.1f" % eps for eps in eps_range], loc="lower right")
plt.xlabel("Number of observations to eps-embed")
plt.ylabel("Minimum number of dimensions")
plt.title("Johnson-Lindenstrauss bounds:\nn_samples vs n_components")
plt.show()
```

执行效果（一）如图 6-3 所示。

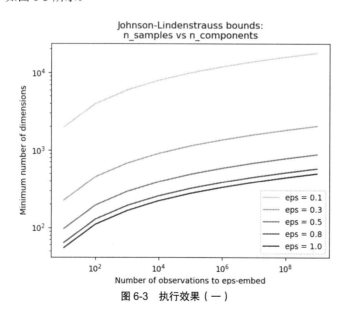

图 6-3　执行效果（一）

（2）对于给定数量的样本，允许失真地增加 eps，以便于大幅减少最小数量的维数 n_components 分量，代码如下：

```
#容许的畸变范围
eps_range = np.linspace(0.01, 0.99, 100)

#要嵌入的样本数(观察值)范围
n_samples_range = np.logspace(2, 6, 5)
colors = plt.cm.Blues(np.linspace(0.3, 1.0, len(n_samples_range)))

plt.figure()
for n_samples, color in zip(n_samples_range, colors):
    min_n_components = johnson_lindenstrauss_min_dim(n_samples, eps=eps_range)
    plt.semilogy(eps_range, min_n_components, color=color)

plt.legend(["n_samples = %d" % n for n in n_samples_range], loc="upper right")
plt.xlabel("Distortion eps")
plt.ylabel("Minimum number of dimensions")
plt.title("Johnson-Lindenstrauss bounds:\nn_components vs eps")
plt.show()
```

执行效果（二）如图 6-4 所示。

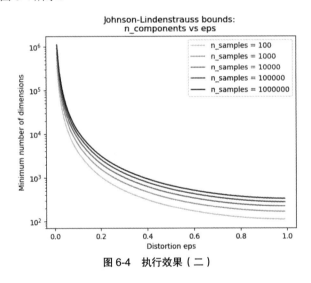

图 6-4　执行效果（二）

（3）验证。

在 20 个新闻组文本文档（TF-IDF 词频）数据集或数字数据集验证了上述界限。

- 对于 20 个新闻组数据集，大约 500 个具有 100k 特征的文档使用稀疏随机矩阵，投影到较小的欧几里得空间，目标维数具有不同的值 n_components。
- 对于数字数据集，将500张手写数字图片的一些8×8灰度像素数据随机投影到各种更大n_components（维度的空间）中。

默认数据集是 20 个新闻组数据集，要在数字数据集上运行本实例，请将 --use-digits-dataset 命令行参数传递给此脚本。

```
if '--use-digits-dataset' in sys.argv:
    data = load_digits().data[:500]
else:
    data = fetch_20newsgroups_vectorized().data[:500]
```

对于每个 n_components 值，会绘制：
- 以原始空间和投影空间中的成对距离分别作为 x 轴和 y 轴的样本对的二维分布。
- 这些距离之比的一维直方图（投影/原始）。

代码如下：

```
n_samples, n_features = data.shape
print("Embedding %d samples with dim %d using various random projections"
      % (n_samples, n_features))

n_components_range = np.array([300, 1000, 10000])
dists = euclidean_distances(data, squared=True).ravel()

#仅选择不相同的样本对
nonzero = dists != 0
dists = dists[nonzero]

for n_components in n_components_range:
    t0 = time()
    rp = SparseRandomProjection(n_components=n_components)
    projected_data = rp.fit_transform(data)
    print("Projected %d samples from %d to %d in %0.3fs"
          % (n_samples, n_features, n_components, time() - t0))
    if hasattr(rp, 'components_'):
        n_bytes = rp.components_.data.nbytes
        n_bytes += rp.components_.indices.nbytes
        print("Random matrix with size: %0.3fMB" % (n_bytes / 1e6))

    projected_dists = euclidean_distances(
        projected_data, squared=True).ravel()[nonzero]

    plt.figure()
    min_dist = min(projected_dists.min(), dists.min())
    max_dist = max(projected_dists.max(), dists.max())
    plt.hexbin(dists, projected_dists, gridsize=100, cmap=plt.cm.PuBu,
               extent=[min_dist, max_dist, min_dist, max_dist])
    plt.xlabel("Pairwise squared distances in original space")
    plt.ylabel("Pairwise squared distances in projected space")
    plt.title("Pairwise distances distribution for n_components=%d" %
              n_components)
    cb = plt.colorbar()
```

```
cb.set_label('Sample pairs counts')
rates = projected_dists / dists
print("Mean distances rate: %0.2f (%0.2f)"
      % (np.mean(rates), np.std(rates)))

plt.figure()
plt.hist(rates, bins=50, range=(0., 2.), edgecolor='k', **density_param)
plt.xlabel("Squared distances rate: projected / original")
plt.ylabel("Distribution of samples pairs")
plt.title("Histogram of pairwise distance rates for n_components=%d" %
          n_components)

#计算eps的预期值,并将其作为垂直线/区域添加到上一个绘图中

plt.show()
```

执行效果（三）如图 6-5 所示。

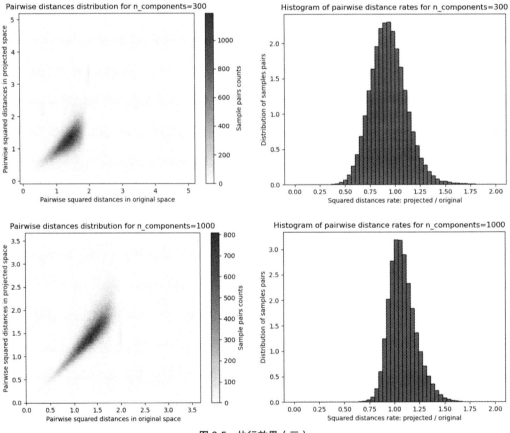

图 6-5　执行效果（三）

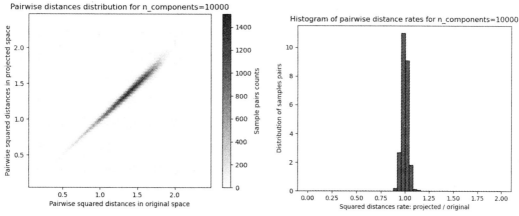

图6-5 执行效果(三)(续)

通过上述执行效果我们可以看到,对于低值n_components的分布很宽,有许多扭曲的对和偏斜的分布(由于最左侧被限制为0,所以距离总是正的),而对于较大的n_components值,失真得到控制并且随机投影很好地保留了距离。

第 7 章
实现大数据计算

在开发机器学习程序的过程中,经常需要计算数据,特别是经常需要计算大数据。在 Scikit-Learn 应用中,提供了专用的库来实现计算大数据的功能。在本章的内容中,将详细介绍使用 Scikit-Learn 计算大数据的知识,为读者步入本书后面知识的学习打下基础。

7.1 计算扩展策略

在 Scikit-Learn 应用程序中，可以考虑使用扩展实现大数据计算功能。

7.1.1 使用外核学习实例进行拓展

扫码观看本节视频讲解

外核（外部存储器）学习是一种用于学习那些无法装进计算机主存储（RAM）的数据的技术，描述了一种为了实现这一目的而设计的系统。
- 一种用流来传输实例的方式。
- 一种从实例中提取特征的方法。
- 增量式算法。

（1）流式实例。

流是从硬盘、数据库、网络流等文件中产生实例，经常被称为输入流。

（2）提取特征

提取特征是 Scikit-Learn 支持的不同特征提取方法 <feature_extraction> 中的任何相关的方法。然而，当处理那些需要矢量化并且特征或值的集合预先不知道的时候，提取特征就变得十分重要了。一个好的例子是文本分类，在训练期间很可能会发现未知的项。从应用的角度上来看，如果在数据上进行多次通过是合理的，则可以使用有状态的向量化器。否则，可以通过使用无状态特征提取器来提高难度。目前，这样做的首选方法是使用所谓的"哈希技巧"，在 sklearn.feature_extraction.FeatureHasher 中，其中有分类变量的表示为 Python 列表或 sklearn.feature_extraction.text.HashingVectorizer 文本文档。

（3）增量学习。

在 Scikit-Learn 中有许多选择，虽然不是所有的算法都能够实现增量学习（即不能一次性看到所有的实例），但是所有的 partial_fit 类 API 估计器都可以作为候选方法实现增量学习功能。实际上，从小批量的实例（有时称为"在线学习"）逐渐学习的能力是外核学习的关键，因为它能保证在任何给定的时间内只有少量的实例在主存储中，选择适合小批量的尺寸来平衡相关性和内存占用可能涉及一些调整。

下面是针对不同任务的增量估算器的列表。

（1）Classification（分类）。
- sklearn.naive_bayes.MultinomialNB。
- sklearn.naive_bayes.BernoulliNB。
- sklearn.linear_model.Perceptron。
- sklearn.linear_model.SGDClassifier。
- sklearn.linear_model.PassiveAggressiveClassifier。
- sklearn.neural_network.MLPClassifier。

（2）Regression（回归）。
- sklearn.linear_model.SGDRegressor。
- sklearn.linear_model.PassiveAggressiveRegressor。

- sklearn.neural_network.MLPRegressor。

（3）Clustering（聚类）。
- sklearn.cluster.MiniBatchKMeans。
- sklearn.cluster.Birch。

（4）Decomposition / feature Extraction（分解 / 特征提取）。
- sklearn.decomposition.MiniBatchDictionaryLearning。
- sklearn.decomposition.IncrementalPCA。
- sklearn.decomposition.LatentDirichletAllocation。

（5）Preprocessing（预处理）。
- sklearn.preprocessing.StandardScaler。
- sklearn.preprocessing.MinMaxScaler。
- sklearn.preprocessing.MaxAbsScaler。

对于分类来说有一点要注意，虽然无状态特征提取程序可能能够应对新的、未知的属性，但是增量学习者本身可能无法应对新的、未知的目标类。在这种情况下，必须使用"classes= 参数"的形式将所有可能的类传递给第一个 partial_fit() 函数调用。

在选择合适的算法时需要考虑的另一个方面是，所有这些算法随着时间的推移不会给每个样例相同的重要性。比如，Perceptron 对错误标签的例子仍然是敏感的，即使经过多次的样例训练，而 SGD* 和 PassiveAggressive* 族之类能够实现更好的健壮性。相反，对于后面传入的数据流，算法的学习速率随着时间不断降低，后面两个算法对于那些显著差异的样本和标注正确的样本倾向给予很少的重视。

7.1.2 使用外核方法进行分类

请看下面的实例文件 fen.py，功能是使用 Scikit-Learn 核外方法进行分类处理。在本实例中使用了一个支持 partial_fit 方法的分类器，为了保证特征空间随着时间的推移保持不变，使用 HashingVectorizer 将每个示例投影到相同的特征空间中。这在文本分类应用中特别有用，在其中每个批次中可能会出现新的特征（词）。实例文件 fen.py 的具体实现流程如下所示。

（1）加载使用 Reuters-21578 数据集，这是由 UCI ML 存储库提供的，在首次运行时会自动下载并解压缩这个数据集。代码如下：

```
class ReutersParser(HTMLParser):
    """实用程序类来解析SGML文件并一次生成一个文档."""

    def __init__(self, encoding='latin-1'):
        HTMLParser.__init__(self)
        self._reset()
        self.encoding = encoding

    def handle_starttag(self, tag, attrs):
        method = 'start_' + tag
        getattr(self, method, lambda x: None)(attrs)
```

```python
    def handle_endtag(self, tag):
        method = 'end_' + tag
        getattr(self, method, lambda: None)()

    def _reset(self):
        self.in_title = 0
        self.in_body = 0
        self.in_topics = 0
        self.in_topic_d = 0
        self.title = ""
        self.body = ""
        self.topics = []
        self.topic_d = ""

    def parse(self, fd):
        self.docs = []
        for chunk in fd:
            self.feed(chunk.decode(self.encoding))
            for doc in self.docs:
                yield doc
            self.docs = []
        self.close()

    def handle_data(self, data):
        if self.in_body:
            self.body += data
        elif self.in_title:
            self.title += data
        elif self.in_topic_d:
            self.topic_d += data

    def start_reuters(self, attributes):
        pass

    def end_reuters(self):
        self.body = re.sub(r'\s+', r' ', self.body)
        self.docs.append({'title': self.title,
                          'body': self.body,
                          'topics': self.topics})
        self._reset()

    def start_title(self, attributes):
        self.in_title = 1

    def end_title(self):
        self.in_title = 0

    def start_body(self, attributes):
```

```python
        self.in_body = 1

    def end_body(self):
        self.in_body = 0

    def start_topics(self, attributes):
        self.in_topics = 1

    def end_topics(self):
        self.in_topics = 0

    def start_d(self, attributes):
        self.in_topic_d = 1

    def end_d(self):
        self.in_topic_d = 0
        self.topics.append(self.topic_d)
        self.topic_d = ""


def stream_reuters_documents(data_path=None):
    """迭代路透社数据集的文档.

路透社档案将自动下载和解压缩,如果

"data\u path"目录不存在.

文档表示为带"body"(str)的字典, (str),
    'title' (str), 'topics' (list(str)) keys.

    """

    DOWNLOAD_URL = ('http://archive.ics.uci.edu/ml/machine-learning-databases/'
                    'reuters21578-mld/reuters21578.tar.gz')
    ARCHIVE_FILENAME = 'reuters21578.tar.gz'

    if data_path is None:
        data_path = os.path.join(get_data_home(), "reuters")
    if not os.path.exists(data_path):
        """Download the dataset."""
        print("downloading dataset (once and for all) into %s" %
              data_path)
        os.mkdir(data_path)

        def progress(blocknum, bs, size):
            total_sz_mb = '%.2f MB' % (size / 1e6)
            current_sz_mb = '%.2f MB' % ((blocknum * bs) / 1e6)
            if _not_in_sphinx():
                sys.stdout.write(
```

```
                    '\rdownloaded %s / %s' % (current_sz_mb, total_sz_mb))

        archive_path = os.path.join(data_path, ARCHIVE_FILENAME)
        urlretrieve(DOWNLOAD_URL, filename=archive_path,
                    reporthook=progress)
        if _not_in_sphinx():
            sys.stdout.write('\r')
        print("untarring Reuters dataset...")
        tarfile.open(archive_path, 'r:gz').extractall(data_path)
        print("done.")

    parser = ReutersParser()
    for filename in glob(os.path.join(data_path, "*.sgm")):
        for doc in parser.parse(open(filename, 'rb')):
            yield doc
```

(2) 创建矢量化器并将特征数量限制为合理的最大值,代码如下:

```
vectorizer = HashingVectorizer(decode_error='ignore', n_features=2 ** 18,
                               alternate_sign=False)

#对已解析的Reuters SGML文件执行迭代器
data_stream = stream_reuters_documents()

#我们学习"acq"类和所有其他类之间的二进制分类

#"acq"之所以被选中,是因为它在路透社的文件中或多或少是均匀分布的

#对于其他数据集,应该注意创建一个测试集,其中包含一部分真实的正实例
all_classes = np.array([0, 1])
positive_class = 'acq'

#下面是一些支持"partial_fit"方法的分类器
partial_fit_classifiers = {
    'SGD': SGDClassifier(max_iter=5),
    'Perceptron': Perceptron(),
    'NB Multinomial': MultinomialNB(alpha=0.01),
    'Passive-Aggressive': PassiveAggressiveClassifier(),
}


def get_minibatch(doc_iter, size, pos_class=positive_class):
    """提取实例的最小批次数量,返回一个元组对象X_text,y



    """
    data = [('{title}\n\n{body}'.format(**doc), pos_class in doc['topics'])
            for doc in itertools.islice(doc_iter, size)
```

```python
                   if doc['topics']]
    if not len(data):
        return np.asarray([], dtype=int), np.asarray([], dtype=int)
    X_text, y = zip(*data)
    return X_text, np.asarray(y, dtype=int)


def iter_minibatches(doc_iter, minibatch_size):
    """小批量发生器."""
    X_text, y = get_minibatch(doc_iter, minibatch_size)
    while len(X_text):
        yield X_text, y
        X_text, y = get_minibatch(doc_iter, minibatch_size)


#测试数据统计
test_stats = {'n_test': 0, 'n_test_pos': 0}

#首先我们举几个例子来估计准确度
n_test_documents = 1000
tick = time.time()
X_test_text, y_test = get_minibatch(data_stream, 1000)
parsing_time = time.time() - tick
tick = time.time()
X_test = vectorizer.transform(X_test_text)
vectorizing_time = time.time() - tick
test_stats['n_test'] += len(y_test)
test_stats['n_test_pos'] += sum(y_test)
print("Test set is %d documents (%d positive)" % (len(y_test), sum(y_test)))


def progress(cls_name, stats):
    """报告进度信息,返回字符串."""
    duration = time.time() - stats['t0']
    s = "%20s classifier : \t" % cls_name
    s += "%(n_train)6d train docs (%(n_train_pos)6d positive) " % stats
    s += "%(n_test)6d test docs (%(n_test_pos)6d positive) " % test_stats
    s += "accuracy: %(accuracy).3f " % stats
    s += "in %.2fs (%5d docs/s)" % (duration, stats['n_train'] / duration)
    return s


cls_stats = {}

for cls_name in partial_fit_classifiers:
    stats = {'n_train': 0, 'n_train_pos': 0,
             'accuracy': 0.0, 'accuracy_history': [(0, 0)], 't0': time.time(),
             'runtime_history': [(0, 0)], 'total_fit_time': 0.0}
    cls_stats[cls_name] = stats
```

```
get_minibatch(data_stream, n_test_documents)
#丢弃测试集
#我们将为分类器提供小批量的1000个文档,这意味着在任何时候的内存中最多有1000个文档
#文件越小,批处理时部分拟合方法的相对开销越大
minibatch_size = 1000

#创建解析Reuters SGML文件并以流形式迭代文档的数据流
minibatch_iterators = iter_minibatches(data_stream, minibatch_size)
total_vect_time = 0.0

#主循环:迭代小批量的示例
for i, (X_train_text, y_train) in enumerate(minibatch_iterators):

    tick = time.time()
    X_train = vectorizer.transform(X_train_text)
    total_vect_time += time.time() - tick

    for cls_name, cls in partial_fit_classifiers.items():
        tick = time.time()
        #用当前mini-batch中的示例更新估计器
        cls.partial_fit(X_train, y_train, classes=all_classes)

        #累积测试精度统计
        cls_stats[cls_name]['total_fit_time'] += time.time() - tick
        cls_stats[cls_name]['n_train'] += X_train.shape[0]
        cls_stats[cls_name]['n_train_pos'] += sum(y_train)
        tick = time.time()
        cls_stats[cls_name]['accuracy'] = cls.score(X_test, y_test)
        cls_stats[cls_name]['prediction_time'] = time.time() - tick
        acc_history = (cls_stats[cls_name]['accuracy'],
                       cls_stats[cls_name]['n_train'])
        cls_stats[cls_name]['accuracy_history'].append(acc_history)
        run_history = (cls_stats[cls_name]['accuracy'],
                       total_vect_time + cls_stats[cls_name]['total_fit_time'])
        cls_stats[cls_name]['runtime_history'].append(run_history)

        if i % 3 == 0:
            print(progress(cls_name, cls_stats[cls_name]))
    if i % 3 == 0:
        print('\n')
```

此时执行后会输出:

```
downloading dataset (once and for all) into /home/circleci/scikit_learn_data/reuters
untarring Reuters dataset...
done.
Test set is 878 documents (108 positive)
          SGD classifier :         962 train docs (    132 positive)      878 test
```

```
docs (    108 positive) accuracy: 0.933 in 0.79s ( 1214 docs/s)
           Perceptron classifier :         962 train docs (    132 positive)       878 test
docs (    108 positive) accuracy: 0.855 in 0.79s ( 1210 docs/s)
         NB Multinomial classifier :       962 train docs (    132 positive)       878 test
docs (    108 positive) accuracy: 0.877 in 0.81s ( 1193 docs/s)
      Passive-Aggressive classifier :      962 train docs (    132 positive)       878 test
docs (    108 positive) accuracy: 0.900 in 0.81s ( 1190 docs/s)


                  SGD classifier :         3911 train docs (    517 positive)       878 test
docs (    108 positive) accuracy: 0.949 in 2.24s ( 1748 docs/s)
           Perceptron classifier :        3911 train docs (    517 positive)       878 test
docs (    108 positive) accuracy: 0.936 in 2.24s ( 1746 docs/s)
         NB Multinomial classifier :      3911 train docs (    517 positive)       878 test
docs (    108 positive) accuracy: 0.885 in 2.27s ( 1725 docs/s)
      Passive-Aggressive classifier :     3911 train docs (    517 positive)       878 test
docs (    108 positive) accuracy: 0.952 in 2.27s ( 1721 docs/s)


                  SGD classifier :         6821 train docs (    891 positive)       878 test
docs (    108 positive) accuracy: 0.932 in 3.67s ( 1856 docs/s)
           Perceptron classifier :        6821 train docs (    891 positive)       878 test
docs (    108 positive) accuracy: 0.952 in 3.68s ( 1854 docs/s)
         NB Multinomial classifier :      6821 train docs (    891 positive)       878 test
docs (    108 positive) accuracy: 0.900 in 3.69s ( 1849 docs/s)
      Passive-Aggressive classifier :     6821 train docs (    891 positive)       878 test
docs (    108 positive) accuracy: 0.959 in 3.69s ( 1847 docs/s)


                  SGD classifier :         9759 train docs (   1276 positive)       878 test
docs (    108 positive) accuracy: 0.960 in 5.13s ( 1902 docs/s)
           Perceptron classifier :        9759 train docs (   1276 positive)       878 test
docs (    108 positive) accuracy: 0.953 in 5.13s ( 1902 docs/s)
         NB Multinomial classifier :      9759 train docs (   1276 positive)       878 test
docs (    108 positive) accuracy: 0.909 in 5.14s ( 1897 docs/s)
      Passive-Aggressive classifier :     9759 train docs (   1276 positive)       878 test
docs (    108 positive) accuracy: 0.960 in 5.15s ( 1893 docs/s)


                  SGD classifier :        11680 train docs (   1499 positive)       878 test
docs (    108 positive) accuracy: 0.940 in 6.49s ( 1799 docs/s)
           Perceptron classifier :       11680 train docs (   1499 positive)       878 test
docs (    108 positive) accuracy: 0.956 in 6.49s ( 1798 docs/s)
         NB Multinomial classifier :     11680 train docs (   1499 positive)       878 test
docs (    108 positive) accuracy: 0.915 in 6.51s ( 1794 docs/s)
      Passive-Aggressive classifier :    11680 train docs (   1499 positive)       878 test
docs (    108 positive) accuracy: 0.945 in 6.51s ( 1794 docs/s)
```

```
              SGD classifier :         14625 train docs (   1865 positive)    878 test
docs (    108 positive) accuracy: 0.960 in 8.12s ( 1800 docs/s)
       Perceptron classifier :         14625 train docs (   1865 positive)    878 test
docs (    108 positive) accuracy: 0.903 in 8.12s ( 1800 docs/s)
     NB Multinomial classifier :       14625 train docs (   1865 positive)    878 test
docs (    108 positive) accuracy: 0.924 in 8.13s ( 1798 docs/s)
  Passive-Aggressive classifier :      14625 train docs (   1865 positive)    878 test
docs (    108 positive) accuracy: 0.964 in 8.14s ( 1797 docs/s)
              SGD classifier :         17360 train docs (   2179 positive)    878 test
docs (    108 positive) accuracy: 0.941 in 9.59s ( 1809 docs/s)
       Perceptron classifier :         17360 train docs (   2179 positive)    878 test
docs (    108 positive) accuracy: 0.933 in 9.60s ( 1809 docs/s)
     NB Multinomial classifier :       17360 train docs (   2179 positive)    878 test
docs (    108 positive) accuracy: 0.932 in 9.60s ( 1807 docs/s)
  Passive-Aggressive classifier :      17360 train docs (   2179 positive)    878 test
docs (    108 positive) accuracy: 0.960 in 9.61s ( 1807 docs/s)
```

（3）绘制表示分类器的学习曲线，展示在小批量过程中分类精度的演变过程，在前 1000 个样本上测量准确性并作为验证集。为了限制内存消耗，在实例中固定排队的数量，然后再将它们提供给学习器。代码如下：

```
def plot_accuracy(x, y, x_legend):
    """绘图精度与x的函数关系."""
    x = np.array(x)
    y = np.array(y)
    plt.title('Classification accuracy as a function of %s' % x_legend)
    plt.xlabel('%s' % x_legend)
    plt.ylabel('Accuracy')
    plt.grid(True)
    plt.plot(x, y)


rcParams['legend.fontsize'] = 10
cls_names = list(sorted(cls_stats.keys()))

#绘图精度演变
plt.figure()
for _, stats in sorted(cls_stats.items()):
    #用#examples绘图精度演变
    accuracy, n_examples = zip(*stats['accuracy_history'])
    plot_accuracy(n_examples, accuracy, "training examples (#)")
    ax = plt.gca()
    ax.set_ylim((0.8, 1))
plt.legend(cls_names, loc='best')

plt.figure()
for _, stats in sorted(cls_stats.items()):
```

```python
    #运行时绘图精度演化
    accuracy, runtime = zip(*stats['runtime_history'])
    plot_accuracy(runtime, accuracy, 'runtime (s)')
    ax = plt.gca()
    ax.set_ylim((0.8, 1))
plt.legend(cls_names, loc='best')

#绘图拟合时间
plt.figure()
fig = plt.gcf()
cls_runtime = [stats['total_fit_time']
               for cls_name, stats in sorted(cls_stats.items())]

cls_runtime.append(total_vect_time)
cls_names.append('Vectorization')
bar_colors = ['b', 'g', 'r', 'c', 'm', 'y']

ax = plt.subplot(111)
rectangles = plt.bar(range(len(cls_names)), cls_runtime, width=0.5,
                     color=bar_colors)

ax.set_xticks(np.linspace(0, len(cls_names) - 1, len(cls_names)))
ax.set_xticklabels(cls_names, fontsize=10)
ymax = max(cls_runtime) * 1.2
ax.set_ylim((0, ymax))
ax.set_ylabel('runtime (s)')
ax.set_title('Training Times')


def autolabel(rectangles):
    """在矩形上附加一些文本vi自动标签."""
    for rect in rectangles:
        height = rect.get_height()
        ax.text(rect.get_x() + rect.get_width() / 2.,
                1.05 * height, '%.4f' % height,
                ha='center', va='bottom')
        plt.setp(plt.xticks()[1], rotation=30)

autolabel(rectangles)
plt.tight_layout()
plt.show()

#绘图预测时间
plt.figure()
cls_runtime = []
cls_names = list(sorted(cls_stats.keys()))
for cls_name, stats in sorted(cls_stats.items()):
    cls_runtime.append(stats['prediction_time'])
cls_runtime.append(parsing_time)
```

```
cls_names.append('Read/Parse\n+Feat.Extr.')
cls_runtime.append(vectorizing_time)
cls_names.append('Hashing\n+Vect.')

ax = plt.subplot(111)
rectangles = plt.bar(range(len(cls_names)), cls_runtime, width=0.5,
                     color=bar_colors)

ax.set_xticks(np.linspace(0, len(cls_names) - 1, len(cls_names)))
ax.set_xticklabels(cls_names, fontsize=8)
plt.setp(plt.xticks()[1], rotation=30)
ymax = max(cls_runtime) * 1.2
ax.set_ylim((0, ymax))
ax.set_ylabel('runtime (s)')
ax.set_title('Prediction Times (%d instances)' % n_test_documents)
autolabel(rectangles)
plt.tight_layout()
plt.show()
```

fen.py 执行后的效果如图 7-1 所示。

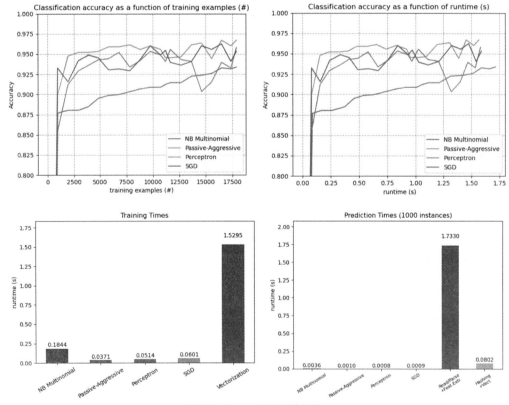

图 7-1　fen.py 执行后的效果

7.2 计算性能

对于某些应用程序来说,估计器的性能(主要是预测时的延迟和吞吐量)至关重要。在 Scikit-Learn 程序中,将预测延迟(prediction latency)作为进行预测所需的经过时间,例如,以微秒(micro-seconds)为单位进行测量。延迟通常被认为是一种分布,运营工程师通常将注意力集中在该分布的给定百分位数(percentile,例如 90 百分位数)上的延迟。

扫码观看本节视频讲解

预测吞吐量(prediction throughput)被定义为软件可以在给定的时间内(例如每秒的预测)可预测的样本数量。

7.2.1 预测延迟

在使用和选择机器学习工具包时,可能遇到的最直接的问题之一是在生产环境中可以预测延迟。影响预测延迟(prediction latency)的主要因素有:
- 特征的数量(number of features);
- 输入数据的表示和稀疏性(input data representation and sparsity);
- 模型复杂性(model complexity);
- 特征提取(feature extraction)。

最后一个主要参数也是在批量或执行一次的时间模式(bulk or one-at-a-time mode)下进行预测的可能性。

1. 批量与原子模式

通常,由于多种原因(分支可预测性、CPU 缓存、线性代数库优化等),通过批量(同时进行多个实例)预测会更加有效。经过对比原子预测延迟和批量预测延迟,发现批量模式的速度总是更快。

如果要对案例的不同 estimators 进行基准测试,可以简单地更改参数 n_features,这样可以估计 prediction latency(预测延迟)的数量级。

2. 配置 Scikit-Learn 以减少验证开销

Scikit-Learn 对数据进行了一些验证,这会增加每次调用 predict 和类似函数的开销,特别是检查这些特征(features)是有限的(不是 NaN 或无限)涉及对数据的完全传递的时候。如果确定数据是可接受的(acceptable),那么可以通过在导入 Scikit-Learn 之前将环境变量 sklearn_assume_finite 设置为非空字符串(non-empty string)来抑制检查有限性,或者使用以下方式在 Python 中配置 sklearn.set_config。为了更加灵活地设置 config_context,允许在指定的上下文中设置此配置:

```
>>>
>>> import sklearn
>>> with sklearn.config_context(assume_finite=True):
...     pass                          #在这里进行学习/预测,减少验证
```

请注意,这将影响在上下文中对 assert_all_finite 的使用。

3. 特征数量的影响

显然，当特征数量增加时，每个示例的内存消耗量也会增加。实际上，对于具有 N 个特征的 M 个实例的矩阵，空间复杂度为 $O(NM)$。

从计算角度来看，这意味着基本操作的数量（例如，线性模型中向量矩阵乘积的乘法）也增加。预测延迟（prediction latency）与特征数量（number of features）的变化，如图 7-2 所示。

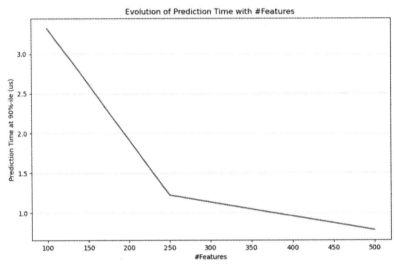

图 7-2　预测延迟与特征数量的变化

总的来说，可以预期预测时间（prediction time）至少会随特征数量（number of features）的变化线性增加 [非线性情况可能会发生，取决于全局内存占用（global memory footprint）和估计（estimator）]。

4. 输入数据表示的影响

Scipy 提供了对存储稀疏数据（storing sparse data）进行优化的稀疏矩阵（sparse matrix）数据结构。稀疏格式（sparse formats）的主要特点是不会存储零，所以如果我们的数据稀疏，那么使用的内存会更少。稀疏（sparse）(CSR 或 CSC) 表示其中的非零值将仅占用一个 32 位整数位置 + 64 位浮点值（floating point）+ 矩阵中每行或列的额外的 32 位。在密集（dense）或稀疏（sparse）线性模型上通过使用稀疏输入可以加速预测，只有非零值特征才会影响点积，从而影响模型预测。因此，如果在 1e6 维空间中有 100 个非零，则只需要 100 次乘法和加法运算，而不是 1e6。

在计算密度结果时可以利用 BLAS 中高度优化的向量操作和多线程技术实现，并且这样可以占用更少的 CPU 高速缓存。因此，稀疏输入（sparse input）表示的稀疏度（sparsity）通常应相当高（10% 非零最大值，要根据硬件进行检查），比在具有多个 CPU 和优化 BLAS 实现的机器上的密集输入（dense input）表示更快。

例如，下面是测试输入稀疏度（sparsity）的示例代码：

```
def sparsity_ratio(X):
    return 1.0 - np.count_nonzero(X) / float(X.shape[0] * X.shape[1])
print("input sparsity ratio:", sparsity_ratio(X))
```

根据经验，可以考虑如果稀疏比（sparsity ratio）大于 90%，可能会从稀疏格式（sparse formats）中受益。

5. 模型复杂度的影响

一般来说，当模型复杂度（model complexity）增加时，预测能力和延迟应该会增加。增加预测能力通常很有意思，但对于许多应用来说，最好不要太多地增加预测延迟。对于 sklearn.linear_model（例如 Lasso、Elastic Net、SGDClassifier/Regresso 等）来说，因为在预测期间应用的决策函数是一样的，所以延迟也应该是等效的。

请看下面的实例文件 fuza.py，功能是演示模型复杂性如何影响预测准确性和计算性能，分别使用了 sklearn.linear_model.stochastic_gradient.SGDClassifier 和惩罚（elastic net penalty）。

本实例将使用两个数据集：
- 用于回归的糖尿病数据集。该数据集由来自糖尿病患者的 10 个测量值组成。任务是预测疾病进展。
- 用于分类的 20 个新闻组文本数据集。该数据集由新闻组帖子组成。任务是预测帖子所写的主题（20 个主题中）。

在本实例中，将对三个不同估计器的复杂性影响进行建模：
- SGDClassifier（用于分类数据）实现随机梯度下降学习。
- NuSVR（用于回归数据）实现 Nu 支持向量回归。
- GradientBoostingRegressor（用于回归数据）以向前阶段的方式构建加性模型。

实例文件 fuza.py 的具体实现流程如下所示。

（1）首先加载两个数据集，使用 fetch_20newsgroups_vectorized 下载 20 个新闻组数据集。这 20 个新闻组数据集是一个稀疏矩阵，而 X 糖尿病数据集是一个 numpy 数组。代码如下：

```
def generate_data(case):
    """Generate regression/classification data."""
    if case == 'regression':
        X, y = datasets.load_diabetes(return_X_y=True)
    elif case == 'classification':
        X, y = datasets.fetch_20newsgroups_vectorized(subset='all',
                                                      return_X_y=True)
    X, y = shuffle(X, y)
    offset = int(X.shape[0] * 0.8)
    X_train, y_train = X[:offset], y[:offset]
    X_test, y_test = X[offset:], y[offset:]

    data = {'X_train': X_train, 'X_test': X_test, 'y_train': y_train,
            'y_test': y_test}
    return data

regression_data = generate_data('regression')
classification_data = generate_data('classification')
```

（2）基准影响。

接下来，可以计算参数对指定估计量的影响。在每一轮中，将使用新的值设置估算器，changing_param 将收集预测时间、预测性能和复杂性，以查看这些变化如何影响估计器。使用 complexity_computer

作为参数传递来计算复杂度。代码如下:

```python
def benchmark_influence(conf):

    #changing_param对MSE和延迟的影响

    prediction_times = []
    prediction_powers = []
    complexities = []
    for param_value in conf['changing_param_values']:
        conf['tuned_params'][conf['changing_param']] = param_value
        estimator = conf['estimator'](**conf['tuned_params'])

        print("Benchmarking %s" % estimator)
        estimator.fit(conf['data']['X_train'], conf['data']['y_train'])
        conf['postfit_hook'](estimator)
        complexity = conf['complexity_computer'](estimator)
        complexities.append(complexity)
        start_time = time.time()
        for _ in range(conf['n_samples']):
            y_pred = estimator.predict(conf['data']['X_test'])
        elapsed_time = (time.time() - start_time) / float(conf['n_samples'])
        prediction_times.append(elapsed_time)
        pred_score = conf['prediction_performance_computer'](
            conf['data']['y_test'], y_pred)
        prediction_powers.append(pred_score)
        print("Complexity: %d | %s: %.4f | Pred. Time: %fs\n" % (
            complexity, conf['prediction_performance_label'], pred_score,
            elapsed_time))
    return prediction_powers, prediction_times, complexities
```

(3) 选择参数。

通过制作一个包含所有必要值的字典来为每个估计器选择参数,changing_param 是在每个估计器中会有所不同的参数的名称。复杂性将由 complexity_label 定义并使用它计算 complexity_computer。另外,请注意,根据估算器类型,需要传递不同的数据。代码如下:

```python
def _count_nonzero_coefficients(estimator):
    a = estimator.coef_.toarray()
    return np.count_nonzero(a)

configurations = [
    {'estimator': SGDClassifier,
     'tuned_params': {'penalty': 'elasticnet', 'alpha': 0.001, 'loss':
                      'modified_huber', 'fit_intercept': True, 'tol': 1e-3},
     'changing_param': 'l1_ratio',
     'changing_param_values': [0.25, 0.5, 0.75, 0.9],
     'complexity_label': 'non_zero coefficients',
     'complexity_computer': _count_nonzero_coefficients,
```

```
     'prediction_performance_computer': hamming_loss,
     'prediction_performance_label': 'Hamming Loss (Misclassification Ratio)',
     'postfit_hook': lambda x: x.sparsify(),
     'data': classification_data,
     'n_samples': 30},
    {'estimator': NuSVR,
     'tuned_params': {'C': 1e3, 'gamma': 2 ** -15},
     'changing_param': 'nu',
     'changing_param_values': [0.1, 0.25, 0.5, 0.75, 0.9],
     'complexity_label': 'n_support_vectors',
     'complexity_computer': lambda x: len(x.support_vectors_),
     'data': regression_data,
     'postfit_hook': lambda x: x,
     'prediction_performance_computer': mean_squared_error,
     'prediction_performance_label': 'MSE',
     'n_samples': 30},
    {'estimator': GradientBoostingRegressor,
     'tuned_params': {'loss': 'ls'},
     'changing_param': 'n_estimators',
     'changing_param_values': [10, 50, 100, 200, 500],
     'complexity_label': 'n_trees',
     'complexity_computer': lambda x: x.n_estimators,
     'data': regression_data,
     'postfit_hook': lambda x: x,
     'prediction_performance_computer': mean_squared_error,
     'prediction_performance_label': 'MSE',
     'n_samples': 30},
]
```

（4）运行代码并绘制结果。

在前面定义了运行基准测试所需的所有函数，现在将遍历我们之前定义的不同配置。然后可以分析从基准测试中获得的图：虽然 L1SGD 分类器中的惩罚减少了预测误差，但是导致了训练时间的增加。我们可以对训练时间进行类似的分析，训练时间随着使用 Nu-SVR 的支持向量的数量而增加。然而，我们观察到有一个最佳数量的支持向量可以减少预测误差。事实上，如果支持向量太少会导致模型欠拟合，而支持向量太多会导致模型过拟合。对于梯度提升模型来说，可以得出完全相同的结论。与 Nu-SVR 的唯一区别是，在集成中拥有过多的树并不那么有害。代码如下：

```
def plot_influence(conf, mse_values, prediction_times, complexities):
    """
    Plot influence of model complexity on both accuracy and latency.
    """

    fig = plt.figure()
    fig.subplots_adjust(right=0.75)

    #第1个折线数据 (prediction error)
    ax1 = fig.add_subplot(111)
```

```python
    line1 = ax1.plot(complexities, mse_values, c='tab:blue', ls='-')[0]
    ax1.set_xlabel('Model Complexity (%s)' % conf['complexity_label'])
    y1_label = conf['prediction_performance_label']
    ax1.set_ylabel(y1_label)

    ax1.spines['left'].set_color(line1.get_color())
    ax1.yaxis.label.set_color(line1.get_color())
    ax1.tick_params(axis='y', colors=line1.get_color())

    #第2个折线数据 (latency)
    ax2 = fig.add_subplot(111, sharex=ax1, frameon=False)
    line2 = ax2.plot(complexities, prediction_times, c='tab:orange', ls='-')[0]
    ax2.yaxis.tick_right()
    ax2.yaxis.set_label_position("right")
    y2_label = "Time (s)"
    ax2.set_ylabel(y2_label)
    ax1.spines['right'].set_color(line2.get_color())
    ax2.yaxis.label.set_color(line2.get_color())
    ax2.tick_params(axis='y', colors=line2.get_color())

    plt.legend((line1, line2), ("prediction error", "latency"),
               loc='upper right')

    plt.title("Influence of varying '%s' on %s" % (conf['changing_param'],
                                                   conf['estimator'].__name__))


for conf in configurations:
    prediction_performances, prediction_times, complexities = \
        benchmark_influence(conf)
    plot_influence(conf, prediction_performances, prediction_times,
                   complexities)
plt.show()
```

请看执行效果，正则化强度（regularization strength）由参数 alpha 全局控制，一个足够高的 alpha 可以增加 elasticnet 的 l1_ratio 参数，以在模型参数中执行各种稀疏程度。这里的 higher sparsity（较高稀疏度）被解释为较少的模型复杂度（less model complexity），因为我们需要较少的系数充分描述它。当然，稀疏性（sparsity）会随着稀疏点积产生的大致时间与非零系数的数目成比例地影响预测时间（prediction time）。en 模型复杂性如图 7-3 所示。

对于 sklearn.svm 具有非线性内核的算法系列，延迟与支持向量的数量相关（越少越快）。延迟和吞吐量应该渐近地随着 SVC 或 SVR 模型中支持向量的数量线性增长。内核也会影响延迟，因为它用于为每个支持向量计算一次输入向量的投影。参数 nu 表示影响支持 NuSVR 向量的数量。NuSVR 模型的复杂性，如图 7-4 所示。

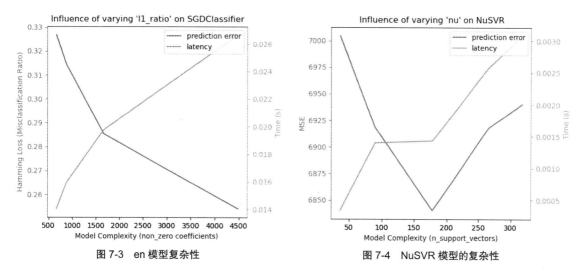

图 7-3 en 模型复杂性

图 7-4 NuSVR 模型的复杂性

对于 sklearn.ensemble 树（例如 RandomForest、GBT、ExtraTrees 等），树的数量及其深度起着最重要的作用。延迟和吞吐量应该与树的数量呈线性关系。在这种情况下，我们直接使用了 n_estimators 参数 GradientBoostingRegressor，如图 7-5 所示。

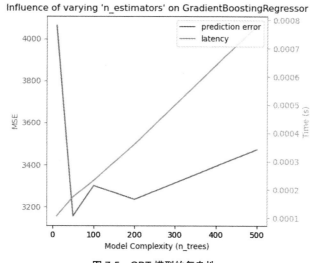

图 7-5 GBT 模型的复杂性

并输出如下结果：

```
Benchmarking SGDClassifier(alpha=0.001, l1_ratio=0.25, loss='modified_huber',
            penalty='elasticnet')
Complexity: 4482 | Hamming Loss (Misclassification Ratio): 0.2541 | Pred. Time: 0.026631s
```

```
Benchmarking SGDClassifier(alpha=0.001, l1_ratio=0.5, loss='modified_huber',
                penalty='elasticnet')
Complexity: 1668 | Hamming Loss (Misclassification Ratio): 0.2854 | Pred. Time: 0.019753s

Benchmarking SGDClassifier(alpha=0.001, l1_ratio=0.75, loss='modified_huber',
                penalty='elasticnet')
Complexity: 874 | Hamming Loss (Misclassification Ratio): 0.3143 | Pred. Time: 0.015934s

Benchmarking SGDClassifier(alpha=0.001, l1_ratio=0.9, loss='modified_huber',
                penalty='elasticnet')
Complexity: 663 | Hamming Loss (Misclassification Ratio): 0.3268 | Pred. Time: 0.014032s

Benchmarking NuSVR(C=1000.0, gamma=3.0517578125e-05, nu=0.1)
Complexity: 36 | MSE: 7004.5333 | Pred. Time: 0.000335s

Benchmarking NuSVR(C=1000.0, gamma=3.0517578125e-05, nu=0.25)
Complexity: 90 | MSE: 6918.2577 | Pred. Time: 0.001395s

Benchmarking NuSVR(C=1000.0, gamma=3.0517578125e-05)
Complexity: 178 | MSE: 6840.2763 | Pred. Time: 0.001427s

Benchmarking NuSVR(C=1000.0, gamma=3.0517578125e-05, nu=0.75)
Complexity: 266 | MSE: 6918.2492 | Pred. Time: 0.002575s

Benchmarking NuSVR(C=1000.0, gamma=3.0517578125e-05, nu=0.9)
Complexity: 318 | MSE: 6940.2899 | Pred. Time: 0.003072s

Benchmarking GradientBoostingRegressor(n_estimators=10)
Complexity: 10 | MSE: 4062.4219 | Pred. Time: 0.000106s

Benchmarking GradientBoostingRegressor(n_estimators=50)
Complexity: 50 | MSE: 3156.4420 | Pred. Time: 0.000173s

Benchmarking GradientBoostingRegressor()
Complexity: 100 | MSE: 3301.5938 | Pred. Time: 0.000232s

Benchmarking GradientBoostingRegressor(n_estimators=200)
Complexity: 200 | MSE: 3235.9376 | Pred. Time: 0.000360s

Benchmarking GradientBoostingRegressor(n_estimators=500)
Complexity: 500 | MSE: 3473.6361 | Pred. Time: 0.000780s
```

通过上述执行效果可以看出：
- 更复杂（或更具表现力）的模型将需要更长的训练时间。
- 更复杂的模型并不能保证减少预测误差。

这些方面与模型泛化和避免模型欠拟合或过拟合有关。

7.2.2 预测吞吐量

考量生产系统大小的另一个重要指标是吞吐量（throughput），即在一定时间内可以做出的预测数量。请看下面的实例文件 tun.py，是基于预测延迟的基准测试，针对合成数据的多个估计器预测吞吐量。目标是测量在批量或原子（即一个接一个）模式下进行预测时可以预期的延迟。

```python
def bulk_benchmark_estimator(estimator, X_test, n_bulk_repeats, verbose):
    """测量整个输入的运行时预测."""
    n_instances = X_test.shape[0]
    runtimes = np.zeros(n_bulk_repeats, dtype=float)
    for i in range(n_bulk_repeats):
        start = time.time()
        estimator.predict(X_test)
        runtimes[i] = time.time() - start
    runtimes = np.array(list(map(lambda x: x / float(n_instances), runtimes)))
    if verbose:
        print("bulk_benchmark runtimes:", min(runtimes), np.percentile(
            runtimes, 50), max(runtimes))
    return runtimes


def benchmark_estimator(estimator, X_test, n_bulk_repeats=30, verbose=False):
    """
    在原子模式和大容量模式下测量预测的运行时
    """
    atomic_runtimes = atomic_benchmark_estimator(estimator, X_test, verbose)
    bulk_runtimes = bulk_benchmark_estimator(estimator, X_test, n_bulk_repeats,
                                              verbose)
    return atomic_runtimes, bulk_runtimes


def generate_dataset(n_train, n_test, n_features, noise=0.1, verbose=False):
    """生成具有给定参数的回归数据集."""
    if verbose:
        print("generating dataset...")

    X, y, coef = make_regression(n_samples=n_train + n_test,
                                  n_features=n_features, noise=noise, coef=True)

    random_seed = 13
    X_train, X_test, y_train, y_test = train_test_split(
        X, y, train_size=n_train, test_size=n_test, random_state=random_seed)
    X_train, y_train = shuffle(X_train, y_train, random_state=random_seed)

    X_scaler = StandardScaler()
    X_train = X_scaler.fit_transform(X_train)
    X_test = X_scaler.transform(X_test)
```

```python
    y_scaler = StandardScaler()
    y_train = y_scaler.fit_transform(y_train[:, None])[:, 0]
    y_test = y_scaler.transform(y_test[:, None])[:, 0]

    gc.collect()
    if verbose:
        print("ok")
    return X_train, y_train, X_test, y_test


def boxplot_runtimes(runtimes, pred_type, configuration):
    """
    用预测运行时的方框图绘制一个新的"图"
    """

    fig, ax1 = plt.subplots(figsize=(10, 6))
    bp = plt.boxplot(runtimes, )

    cls_infos = ['%s\n(%d %s)' % (estimator_conf['name'],
                                  estimator_conf['complexity_computer'](
                                      estimator_conf['instance']),
                                  estimator_conf['complexity_label']) for
                 estimator_conf in configuration['estimators']]
    plt.setp(ax1, xticklabels=cls_infos)
    plt.setp(bp['boxes'], color='black')
    plt.setp(bp['whiskers'], color='black')
    plt.setp(bp['fliers'], color='red', marker='+')

    ax1.yaxis.grid(True, linestyle='-', which='major', color='lightgrey',
                   alpha=0.5)

    ax1.set_axisbelow(True)
    ax1.set_title('Prediction Time per Instance - %s, %d feats.' % (
        pred_type.capitalize(),
        configuration['n_features']))
    ax1.set_ylabel('Prediction Time (us)')

    plt.show()

def benchmark(configuration):
    """运行整个基准测试"""
    X_train, y_train, X_test, y_test = generate_dataset(
        configuration['n_train'], configuration['n_test'],
        configuration['n_features'])

    stats = {}
    for estimator_conf in configuration['estimators']:
```

```python
        print("Benchmarking", estimator_conf['instance'])
        estimator_conf['instance'].fit(X_train, y_train)
        gc.collect()
        a, b = benchmark_estimator(estimator_conf['instance'], X_test)
        stats[estimator_conf['name']] = {'atomic': a, 'bulk': b}

    cls_names = [estimator_conf['name'] for estimator_conf in configuration[
        'estimators']]
    runtimes = [1e6 * stats[clf_name]['atomic'] for clf_name in cls_names]
    boxplot_runtimes(runtimes, 'atomic', configuration)
    runtimes = [1e6 * stats[clf_name]['bulk'] for clf_name in cls_names]
    boxplot_runtimes(runtimes, 'bulk (%d)' % configuration['n_test'],
                     configuration)


def n_feature_influence(estimators, n_train, n_test, n_features, percentile):
    """
    估计特征数量对预测时间的影响.
    """
    percentiles = defaultdict(defaultdict)
    for n in n_features:
        print("benchmarking with %d features" % n)
        X_train, y_train, X_test, y_test = generate_dataset(n_train, n_test, n)
        for cls_name, estimator in estimators.items():
            estimator.fit(X_train, y_train)
            gc.collect()
            runtimes = bulk_benchmark_estimator(estimator, X_test, 30, False)
            percentiles[cls_name][n] = 1e6 * np.percentile(runtimes,
                                                           percentile)
    return percentiles

def plot_n_features_influence(percentiles, percentile):
    fig, ax1 = plt.subplots(figsize=(10, 6))
    colors = ['r', 'g', 'b']
    for i, cls_name in enumerate(percentiles.keys()):
        x = np.array(sorted([n for n in percentiles[cls_name].keys()]))
        y = np.array([percentiles[cls_name][n] for n in x])
        plt.plot(x, y, color=colors[i], )
    ax1.yaxis.grid(True, linestyle='-', which='major', color='lightgrey',
                   alpha=0.5)
    ax1.set_axisbelow(True)
    ax1.set_title('Evolution of Prediction Time with #Features')
    ax1.set_xlabel('#Features')
    ax1.set_ylabel('Prediction Time at %d%%-ile (us)' % percentile)
    plt.show()


def benchmark_throughputs(configuration, duration_secs=0.1):
    """不同估计器的基准吞吐量."""
```

```python
    X_train, y_train, X_test, y_test = generate_dataset(
        configuration['n_train'], configuration['n_test'],
        configuration['n_features'])
    throughputs = dict()
    for estimator_config in configuration['estimators']:
        estimator_config['instance'].fit(X_train, y_train)
        start_time = time.time()
        n_predictions = 0
        while (time.time() - start_time) < duration_secs:
            estimator_config['instance'].predict(X_test[[0]])
            n_predictions += 1
        throughputs[estimator_config['name']] = n_predictions / duration_secs
    return throughputs


def plot_benchmark_throughput(throughputs, configuration):
    fig, ax = plt.subplots(figsize=(10, 6))
    colors = ['r', 'g', 'b']
    cls_infos = ['%s\n(%d %s)' % (estimator_conf['name'],
                                  estimator_conf['complexity_computer'](
                                      estimator_conf['instance']),
                                  estimator_conf['complexity_label']) for
                 estimator_conf in configuration['estimators']]
    cls_values = [throughputs[estimator_conf['name']] for estimator_conf in
                  configuration['estimators']]
    plt.bar(range(len(throughputs)), cls_values, width=0.5, color=colors)
    ax.set_xticks(np.linspace(0.25, len(throughputs) - 0.75, len(throughputs)))
    ax.set_xticklabels(cls_infos, fontsize=10)
    ymax = max(cls_values) * 1.2
    ax.set_ylim((0, ymax))
    ax.set_ylabel('Throughput (predictions/sec)')
    ax.set_title('Prediction Throughput for different estimators (%d '
                 'features)' % configuration['n_features'])
    plt.show()


# #############################################################################
# Main code

start_time = time.time()

# #############################################################################
#各种回归器的基准批量/原子预测速度
configuration = {
    'n_train': int(1e3),
    'n_test': int(1e2),
    'n_features': int(1e2),
    'estimators': [
        {'name': 'Linear Model',
```

```
                 'instance': SGDRegressor(penalty='elasticnet', alpha=0.01,
                                         l1_ratio=0.25, tol=1e-4),
                 'complexity_label': 'non-zero coefficients',
                 'complexity_computer': lambda clf: np.count_nonzero(clf.coef_)},
                {'name': 'RandomForest',
                 'instance': RandomForestRegressor(),
                 'complexity_label': 'estimators',
                 'complexity_computer': lambda clf: clf.n_estimators},
                {'name': 'SVR',
                 'instance': SVR(kernel='rbf'),
                 'complexity_label': 'support vectors',
                 'complexity_computer': lambda clf: len(clf.support_vectors_)},
            ]
}
benchmark(configuration)

#基准n_features对预测速度的影响
percentile = 90
percentiles = n_feature_influence({'ridge': Ridge()},
                                  configuration['n_train'],
                                  configuration['n_test'],
                                  [100, 250, 500], percentile)
plot_n_features_influence(percentiles, percentile)

#基准吞吐量
throughputs = benchmark_throughputs(configuration)
plot_benchmark_throughput(throughputs, configuration)

stop_time = time.time()
print("example run in %.2fs" % (stop_time - start_time))
```

tun.py 的执行效果如图 7-6 所示，这些图将预测延迟的分布表示为箱线图。

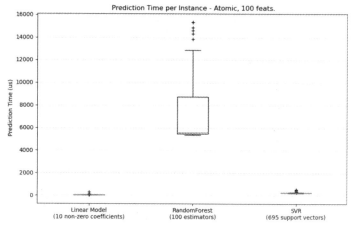

图 7-6　tun.py 的执行效果

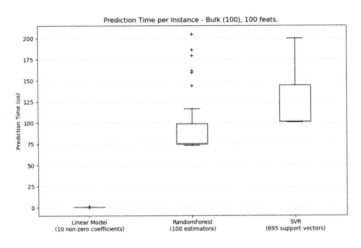

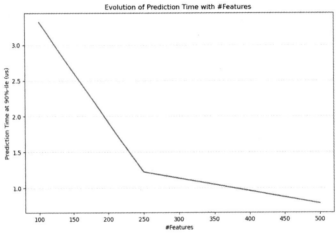

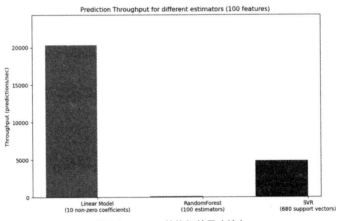

图 7-6 tun.py 的执行效果（续）

第 8 章

英超联赛比分预测系统（Matplotlib+Scikit-Learn+Flask+Pandas）

至此，Scikit-Learn 的基础知识全部介绍完毕。在本章的内容中，将使用 Scikit-Learn 开发一个英超联赛比分预测系统，详细介绍使用 Scikit-Learn 开发大型机器学习程序的知识。

8.1 英超联赛介绍

英格兰足球超级联赛（premier league，简称英超联赛）是英格兰足球总会属下的最高等级职业足球联赛，前身是英格兰足球甲级联赛。英超联赛是"欧洲足球五大联赛"之一，由 20 支球队组成，由超级联盟负责具体运作，赛季结束后积分榜末三位降入英格兰足球冠军联赛。

扫码观看本节视频讲解

英超联赛一直以来被认为是世界上最好的联赛之一，节奏快、竞争激烈、强队众多，是全世界商业运作最成功的联赛与收入最高的足球联赛。

英超联赛采取主客场双循环赛制比赛，每支队伍与各球队对赛两次，主客各一次。从 1995/1996 赛季开始参赛球队由 22 队减至 20 队，每支球队共进行 38 场赛事，主场和客场比赛各有 19 场。每场胜方可得 3 分，平局各得 1 分，负方得 0 分，按各队于联赛所得的积分排列。完成所有赛事后总积分最高的队伍可以夺得联赛冠军，而总积分最低的 3 队球队会降级至英冠联赛。

8.2 系统模块介绍

本系统将基于过去几年的英超球队的战绩数据为基础，然后使用 Scikit-Learn 训练数据并制作模型，根据模型分析预测新赛季英超球队的比赛结果：胜、平、负。本项目的基本模块如图 8-1 所示。

扫码观看本节视频讲解

图 8-1　模块结构

8.3 数据集

本项目用到的数据是从 api-football 获取的，这是一个著名的第三方 API，提供了世界各地足球、篮球等主流体育运动的比赛结果和相关信息。使用 api-football 数据的好处如下。

扫码观看本节视频讲解

(1)可以每天进行 API 调用,使用最新的统计数据和结果刷新数据库,从而使模型能够始终如一地根据最新信息重新训练。

(2)该 API 不仅提供过去的比赛数据,还提供了有关即将到来的比赛信息,这对于进行网络应用程序的预测至关重要。

8.3.1 获取 api-football 密钥

登录 https://www.api-football.com/ 并注册会员,然后在后台页面单击左侧的 Account 按钮,进入 https://dashboard.api-football.com/profile?access 页面,在此页面可以获取足球 API 的密钥(Api Key),如图 8-2 所示。

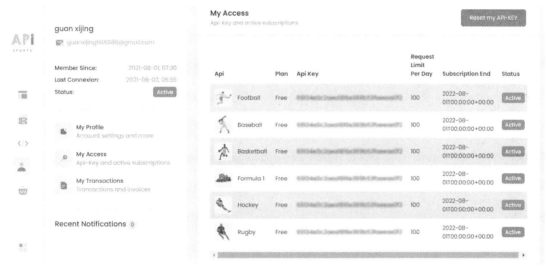

图 8-2 获取密钥

8.3.2 获取数据

编写实例文件 01_api_data_request.py,功能是从 api-football 获取英超联赛的数据,收集 2019—2020 年英超联赛的数据信息,将获取的数据信息保存为 CSV 文件。在 CSV 文件中保存了获取的数据信息,这些数据信息是 JSON 格式的,这些 JSON 格式的数据被加载到 Pandas DataFrame 中,并转换为以下形式的嵌套字典:

```
{team ID: {fixture_id: stats_df}}
```

实例文件 01_api_data_request.py 的具体实现代码如下所示。

```
YEAR = 2021
YEAR_str = str(YEAR)
```

```python
request_league_ids = False
request_fixtures = True
request_missing_game_stats = True

#--------------------------- REQUEST FUNCTIONS ---------------------------
api_key = (open('api_key.txt', mode='r')).read()

def get_api_data(base_url, end_url):
    url = base_url + end_url
    headers = {'X-RapidAPI-Key': api_key}
    res = requests.get(url, headers=headers)
    if res.status_code != 200:
        raise RuntimeError(f'error {res.status_code}')
    res_t = res.text
    return res_t

def slice_api(api_str_output, start_char, end_char):
  e = len(api_str_output) - end_char
  s = start_char
  output = api_str_output[s:e]
  return output

def save_api_output(save_name, jason_data, json_data_path=''):
    writeFile = open(json_data_path + save_name + '.json', 'w')
    writeFile.write(jason_data)
    writeFile.close()

def read_json_as_pd_df(json_data, json_data_path='', orient_def='records'):
    output = pd.read_json(json_data_path + json_data, orient=orient_def)
    return output

#--------------------------- REQUESTING BASIC DATA ---------------------------

base_url = 'https://v2.api-football.com/'


def req_prem_fixtures_id(season_code, year=YEAR_str):
    #请求api获取数据
    premier_league_fixtures_raw = get_api_data(base_url, f'/fixtures/league/{season_code}/')

    #清理数据以准备加载到数据帧中
    premier_league_fixtures_sliced = slice_api(premier_league_fixtures_raw, 33, 2)

    #将提取的数据保存为json文件
    save_api_output(f'{year}_premier_league_fixtures', premier_league_fixtures_sliced,
json_data_path = 'prem_clean_fixtures_and_dataframes/')

    #将json文件作为数据帧加载
```

```python
    premier_league_fixtures_df = read_json_as_pd_df(f'{year}_premier_league_fixtures.json',
json_data_path='prem_clean_fixtures_and_dataframes/')
    return premier_league_fixtures_df

#请求英超联赛的数据,我们将使用此响应获取我们感兴趣的赛季的联赛id
if request_league_ids:
    leagues = premier_league_fixtures_raw = get_api_data(base_url, 'leagues/search/premier_league')

if YEAR == 2019:
    season_id = 524
elif YEAR == 2020:
    season_id = 2790
elif YEAR == 2021:
    season_id = 3456
else:
    print('please lookup season id and specify this as season_id variable')

#使用函数req_prem_fixture_id请求设备列表
if request_fixtures:
    fixtures = req_prem_fixtures_id(season_id, YEAR_str)


def load_prem_fixtures_id(year=YEAR_str):
    premier_league_fixtures_df = read_json_as_pd_df(f'{year}_premier_league_fixtures.json',
json_data_path='prem_clean_fixtures_and_dataframes/')
    return premier_league_fixtures_df

fixtures = load_prem_fixtures_id()

#------------------------ MAKING CLEAN FIXTURE LIST -------------------------

fixtures = pd.read_json(f'prem_clean_fixtures_and_dataframes/{YEAR_str}_premier_league_fixtures.json', orient='records')

#创建干净的历史fixtures列表数据框

for i in fixtures.index:
    x1 = str(fixtures['homeTeam'].iloc[i])[12:14]
    x = int(x1)
    fixtures.at[i, 'HomeTeamID'] = x

for i in fixtures.index:
    x1 = str(fixtures['awayTeam'].iloc[i])[12:14]
    x = int(x1)
    fixtures.at[i, 'AwayTeamID'] = x

for i in fixtures.index:
    x = str(fixtures['event_date'].iloc[i])[:10]
    fixtures.at[i, 'Game Date'] = x
```

```python
    for i in fixtures.index:
        x = str(fixtures['homeTeam'][i]['team_name'])
        fixtures.at[i, 'Home Team'] = x

    for i in fixtures.index:
        x = str(fixtures['awayTeam'][i]['team_name'])
        fixtures.at[i, 'Away Team'] = x

    for i in fixtures.index:
        x = str(fixtures['homeTeam'][i]['logo'])
        fixtures.at[i, 'Home Team Logo'] = x

    for i in fixtures.index:
        x = str(fixtures['awayTeam'][i]['logo'])
        fixtures.at[i, 'Away Team Logo'] = x

    fixtures_clean = pd.DataFrame({'Fixture ID': fixtures['fixture_id'], 'Game Date': fixtures['Game Date'], 'Home Team ID': fixtures['HomeTeamID'], 'Away Team ID': fixtures['AwayTeamID'], 'Home Team Goals': fixtures['goalsHomeTeam'], 'Away Team Goals': fixtures['goalsAwayTeam'], 'Venue': fixtures['venue'], 'Home Team': fixtures['Home Team'], 'Away Team': fixtures['Away Team'], 'Home Team Logo': fixtures['Home Team Logo'], 'Away Team Logo': fixtures['Away Team Logo']})

    fixtures_clean.to_csv(f'prem_clean_fixtures_and_dataframes/{YEAR_str}_premier_league_fixtures_df.csv', index=False)

    #------------------------- STITCHINING CLEAN FIXTURE LIST -------------------------

    #加载2019+2020和2021年的数据,并将两个数据帧缝合在一起
    fixtures_clean_2019_2020 = pd.read_csv('prem_clean_fixtures_and_dataframes/2019_2020_premier_league_fixtures_df.csv')

    fixtures_clean_2021 = pd.read_csv('prem_clean_fixtures_and_dataframes/2021_premier_league_fixtures_df.csv')

    fixtures_clean_combined = pd.concat([fixtures_clean_2019_2020, fixtures_clean_2021])
    fixtures_clean_combined = fixtures_clean_combined.reset_index(drop=True)

    fixtures_clean_combined.to_csv('prem_clean_fixtures_and_dataframes/2019_2020_2021_premier_league_fixtures_df.csv', index=False)

    #------------------------- REQUESTING SPECIFIC STATS -------------------------
    fixtures_clean = pd.read_csv(f'prem_clean_fixtures_and_dataframes/{YEAR_str}_premier_league_fixtures_df.csv')

    def req_prem_stats(start_index, end_index):
        for i in fixtures_clean.index[start_index:end_index]:
```

```python
        if math.isnan(fixtures_clean['Home Team Goals'].iloc[i]) == False:
            fix_id = str(fixtures_clean['Fixture ID'].iloc[i])
            fixture_raw = get_api_data(base_url, '/statistics/fixture/' + fix_id + '/')
            fixture_sliced = slice_api(fixture_raw, 34, 2)
            save_api_output('prem_game_stats_json_files/' + fix_id, fixture_sliced)


#req_prem_stats(288, 300)


#----- AUTOMATING MISSING DATA COLLECTION -----

#搜索现有数据库(prem_game_stats_json_files文件夹),并请求自上次请求数据以来的比赛数据
#列出已收集的json数据
existing_data_raw = listdir('prem_game_stats_json_files/')

#从此列表末尾删除".json"
existing_data = []
for i in existing_data_raw:
    existing_data.append(int(i[:-5]))

#创建一个缺失列表
missing_data = []
for i in fixtures_clean.index:
    fix_id = fixtures_clean['Fixture ID'].iloc[i]
    if fix_id not in existing_data:
        if math.isnan(fixtures_clean['Home Team Goals'].iloc[i]) == False:
            missing_data.append(fix_id)


def req_prem_stats_list(missing_data):
    if len(missing_data) > 100:
        print('This request exceeds 100 request limit and has not been completed')
    else:
        if len(missing_data) > 0:
            print('Data collected for the following fixtures:')
        for i, dat in enumerate(missing_data):
            pause_points = [(i*10)-1 for i in range(10)]
            if i in pause_points:
                print('sleeping for 1 minute - API only allows 10 requests per minute')
                time.sleep(60)
            print(dat)
            fix_id = str(dat)
            fixture_raw = get_api_data(base_url, '/statistics/fixture/' + fix_id + '/')
            fixture_sliced = slice_api(fixture_raw, 34, 2)
            save_api_output('prem_game_stats_json_files/' + fix_id, fixture_sliced)
```

```python
if request_missing_game_stats:
    req_prem_stats_list(missing_data)

# -------------------------------- END --------------------------------------

print('\n', 'Script runtime:', round(((time.time()-start)/60), 2), 'minutes')
print(' ----------------- END ----------------- \n')
```

执行后会创建两个 CSV 文件：2021_premier_league_fixtures.json 和 2021_premier_league_fixtures_df.csv，在这两个文件中保存了获取的 JSON 格式的英超联赛数据信息。

8.3.3 收集最新数据

编写实例文件 02_cleaning_stats_data.py，功能是收集最新的英超联赛数据信息。因为随着时间的推移，英超球队的交锋信息越来越多，通过此文件可以只收集我们目前没有保存的数据信息，然后将这些最新信息添加到本地 CSV 文件中。文件 02_cleaning_stats_data.py 的具体实现代码如下所示。

```python
import pandas as pd
import math
import pickle

#------------------------------- INPUT VARIABLES -------------------------------
#要生成字典的fixtures数据框的名称,以及保存的输出文件(嵌套统计字典)的名称
fixtures_saved_name = '2019_2020_2021_premier_league_fixtures_df.csv'
stats_dict_output_name = '2019_2020_2021_prem_all_stats_dict.txt'

#----------------------------- CREATING DF PER TEAM ----------------------------
#创建一个包含20个球队的嵌套字典,每个团队都有一个值作为另一个字典.在本字典中,我们将有比赛id
 和比赛数据框
fixtures_clean = pd.read_csv(f'prem_clean_fixtures_and_dataframes/{fixtures_saved_name}')

#创建"fixtures_clean"ID索引,我们将使用该索引从该数据帧获取数据,并将其添加到每个fixture stats
 数据帧中
fixtures_clean_ID_index = pd.Index(fixtures_clean['Fixture ID'])

#可以迭代的球队id列表
team_id_list = (fixtures_clean['Home Team ID'].unique()).tolist()

#创建将填充数据的词典
all_stats_dict = {}

#嵌套for循环以创建嵌套字典,第一个键表示球队id,第二个键表示球队比赛id
for team in team_id_list:
    #主队
    team_fixture_list = []
    for i in fixtures_clean.index[:]:
```

```python
        if fixtures_clean['Home Team ID'].iloc[i] == team:
            if math.isnan(fixtures_clean['Home Team Goals'].iloc[i]) == False:
                team_fixture_list.append(fixtures_clean['Fixture ID'].iloc[i])
    all_stats_dict[team] = {}
    for j in team_fixture_list:
        #加载df
        df = pd.read_json('prem_game_stats_json_files/' + str(j) + '.json', orient='values')
        #删除占有中的百分比符号,并将其传递和转换为int
        df['Ball Possession'] = df['Ball Possession'].str.replace('[\%]', '').astype(int)
        df['Passes %'] = df['Passes %'].str.replace('[\%]', '').astype(int)
        #将主客场进球添加到df
        temp_index = fixtures_clean_ID_index.get_loc(j)
        home_goals = fixtures_clean['Home Team Goals'].iloc[temp_index]
        away_goals = fixtures_clean['Away Team Goals'].iloc[temp_index]
        df['Goals'] = [home_goals, away_goals]
        #添加点数据
        if home_goals > away_goals:
            df['Points'] = [2,0]
        elif home_goals == away_goals:
            df['Points'] = [1,1]
        elif home_goals < away_goals:
            df['Points'] = [0,2]
        else:
            df['Points'] = ['nan', 'nan']
        #向df添加主客场标识符
        df['Team Identifier'] = [1,2]
        #添加球队id
        df['Team ID'] = [team, fixtures_clean['Away Team ID'].iloc[temp_index]]
        #添加比赛日期
        gd = fixtures_clean['Game Date'].iloc[temp_index]
        df['Game Date'] = [gd, gd]
        #将此修改的df添加到嵌套字典
        sub_dict_1 = {j:df}
        all_stats_dict[team].update(sub_dict_1)

#客队
team_fixture_list = []
for i in fixtures_clean.index[:]:
    if fixtures_clean['Away Team ID'].iloc[i] == team:
        if math.isnan(fixtures_clean['Away Team Goals'].iloc[i]) == False:
            team_fixture_list.append(fixtures_clean['Fixture ID'].iloc[i])
for j in team_fixture_list:
    #加载df
    df = pd.read_json('prem_game_stats_json_files/' + str(j) + '.json', orient='values')
    #删除占有中的百分比符号,并将其传递和转换为int
    df['Ball Possession'] = df['Ball Possession'].str.replace('[\%]', '').astype(int)
    df['Passes %'] = df['Passes %'].str.replace('[\%]', '').astype(int)
    #将主客场进球添加到df
    temp_index = fixtures_clean_ID_index.get_loc(j)
```

```
            home_goals = fixtures_clean['Home Team Goals'].iloc[temp_index]
            away_goals = fixtures_clean['Away Team Goals'].iloc[temp_index]
            df['Goals'] = [home_goals, away_goals]
            #添加点数据
            if home_goals > away_goals:
                df['Points'] = [2,0]
            elif home_goals == away_goals:
                df['Points'] = [1,1]
            elif home_goals < away_goals:
                df['Points'] = [0,2]
            else:
                df['Points'] = ['nan', 'nan']
            #向df添加主客场标识符
            df['Team Identifier'] = [2,1]
            #添加球队id
            df['Team ID'] = [fixtures_clean['Home Team ID'].iloc[temp_index], team]
            #添加比赛日期
            gd = fixtures_clean['Game Date'].iloc[temp_index]
            df['Game Date'] = [gd, gd]
            #将此修改的df添加到嵌套字典
            sub_dict_1 = {j:df}
            all_stats_dict[team].update(sub_dict_1)

#将生成的字典保存为pickle文件,以便导入到以后的python文件中
with open(f'prem_clean_fixtures_and_dataframes/{stats_dict_output_name}', 'wb') as myFile:
    pickle.dump(all_stats_dict, myFile)
with open(f'prem_clean_fixtures_and_dataframes/{stats_dict_output_name}', 'rb') as myFile:
    loaded_dict_test = pickle.load(myFile)

# --------------------------------- END ---------------------------------
print('\n', 'Script runtime:', round(((time.time()-start)/60), 2), 'minutes')
print(' ---------------- END ---------------- \n')
```

8.4 特征提取和数据可视化

为了尽可能多地利用之前的比赛数据,同时最大限度地减少特征数量,对之前 10 场比赛的数据进行平均以预测即将到来的比赛结果。为了了解某一支球队的表现如何,将他们的平均数据从对手的平均数据中减去,以得到一个差异指标,例如射门次数差异。如果一支球队 number_of_shots_diff = 2,这表示在过去 10 场比赛中平均比对手多射门 2 次。在本实例中我们选择了 7 个 "差异" 特征:

扫码观看本节视频讲解

- 目标差异;
- 射门差;
- 镜头差异;
- 占有差异;

- 通过精度差异；
- 角差；
- 犯规差。

上面的差异描述了单个团队的特征，因此在预测比赛结果时，特征数量翻了一番，达到 14 个。

8.4.1 提取数据

编写实例文件 03_feature_engineering.py，功能是提取本地数据集中的数据，分别提取过去 5 场比赛的数据和过去 10 场比赛的数据。具体实现代码如下所示。

```
#使用"02_cleaning_stats_data.py"生成的已保存嵌套字典的名称,以及已保存的输出文件(stats DataFrame)的名称
stats_dict_saved_name = '2019_2020_2021_prem_all_stats_dict.txt'
df_5_output_name = '2019_2020_2021_prem_df_for_ml_5_v2.txt'
df_10_output_name = '2019_2020_2021_prem_df_for_ml_10_v2.txt'

#---------------------------- FEATURE ENGINEERING ----------------------------
with open(f'prem_clean_fixtures_and_dataframes/{stats_dict_saved_name}', 'rb') as myFile:
    game_stats = pickle.load(myFile)

#创建保存球队id的列表
team_list = []
for key in game_stats.keys():
    team_list.append(key)
team_list.sort()

#创建一个字典,其中球队id为键,fixtureid为值
team_fixture_id_dict = {}
for team in team_list:
    fix_id_list = []
    for key in game_stats[team].keys():
        fix_id_list.append(key)
    fix_id_list.sort()
    sub_dict = {team:fix_id_list}
    team_fixture_id_dict.update(sub_dict)

#按时间顺序排列fixtureid
for team in team_fixture_id_dict:
    team_fixture_id_dict[team].sort()

#使用上面创建的dict在给定球队id键的fixture id列表上进行迭代。注意,过去数据平均的场次数
#一个大的数字将平滑表现,而一个小的数字将导致预测严重依赖于最近的形式。这值得在ml模型构建阶段进行测试
#5场比赛均线
df_ml_5 = average_stats_df(5, team_list, team_fixture_id_dict, game_stats)
```

```
#10场比赛均线
df_ml_10 = average_stats_df(10, team_list, team_fixture_id_dict, game_stats)

#创建并保存具有5场比赛滑动平均值的ml数据帧
df_for_ml_5_v2 = creating_ml_df(df_ml_5)
with open(f'prem_clean_fixtures_and_dataframes/{df_5_output_name}', 'wb') as myFile:
    pickle.dump(df_for_ml_5_v2, myFile)

#使用10场比赛的滑动平均值创建和保存ml数据帧
df_for_ml_10_v2 = creating_ml_df(df_ml_10)
with open(f'prem_clean_fixtures_and_dataframes/{df_10_output_name}', 'wb') as myFile:
    pickle.dump(df_for_ml_10_v2, myFile)

# --------------------------------- END ---------------------------------
print('\n', 'Script runtime:', round(((time.time()-start)/60), 2), 'minutes')
print(' ---------------- END ---------------- \n')
```

执行后会创建两个文件 2019_2020_2021_prem_df_for_ml_10_v2.txt 和 2019_2020_2021_prem_df_for_ml_5_v2.txt，在里面分别保存了过去 5 场比赛和过去 10 场比赛的数据信息。

8.4.2 数据可视化

编写实例文件 04_feature_engineering_data_visualisation.py，功能是根据过去 5 场比赛和过去 10 场比赛的数据信息绘制可视化图表。主要实现代码如下所示。

```
plt.rcParams['font.sans-serif'] = 'SimHei'
plt.rcParams['axes.unicode_minus'] = False  ## 设置正常显示符号

df_5_saved_name = '2019_2020_prem_df_for_ml_5_v2.txt'
df_10_saved_name = '2019_2020_prem_df_for_ml_10_v2.txt'

save_df_10_fig = False
save_df_5_fig = False

colourbar = 'winter'

#其中,o表示绘制球队失败图,1将绘制平局图,2将绘制团队胜利图
plot_results = [1]

#------------------------ PRE-ML DATA VISUALISATION ------------------------
with open(f'prem_clean_fixtures_and_dataframes/{df_5_saved_name}', 'rb') as myFile:
    df_ml_5 = pickle.load(myFile)

with open(f'prem_clean_fixtures_and_dataframes/{df_10_saved_name}', 'rb') as myFile:
    df_ml_10 = pickle.load(myFile)

#---------- DATA PREP ----------
```

```python
#删除了不希望绘制的比赛结果，因为这些结果受plot_results的控制

for i in range(0, len(df_ml_10)):
    if df_ml_10['Team Result Indicator'].loc[i] in plot_results:
        continue
    else:
        df_ml_10 = df_ml_10.drop([i], axis=0)

df_ml_10 = df_ml_10.reset_index(drop=True)

for i in range(0, len(df_ml_5)):
    if df_ml_5['Team Result Indicator'].loc[i] in plot_results:
        continue
    else:
        df_ml_5 = df_ml_5.drop([i], axis=0)

df_ml_5 = df_ml_5.reset_index(drop=True)
#--------------------------------- FIGURE 1 ---------------------------------
#figure 1 - 设置包装器
fig, ((ax1, ax2, ax3),(ax4, ax5, ax6)) = plt.subplots(ncols=3,nrows=2,
                                                      figsize=(18,12))

fig.suptitle('Data Averaged Over 10 Games', y=0.99, fontsize=16, fontweight='bold');

#绘制6个数字
scat1 = ax1.scatter(df_ml_10['Team Av Shots Diff'],
                    df_ml_10['Opponent Av Shots Diff'],
                    c=df_ml_10['Team Result Indicator'],
                    cmap = colourbar);

scat2 = ax2.scatter(df_ml_10['Team Av Shots Inside Box Diff'],
                    df_ml_10['Opponent Av Shots Inside Box Diff'],
                    c=df_ml_10['Team Result Indicator'],
                    cmap = colourbar);

scat3 = ax3.scatter(df_ml_10['Team Av Fouls Diff'],
                    df_ml_10['Opponent Av Fouls Diff'],
                    c=df_ml_10['Team Result Indicator'],
                    cmap = colourbar);

scat4 = ax4.scatter(df_ml_10['Team Av Corners Diff'],
                    df_ml_10['Opponent Av Corners Diff'],
                    c=df_ml_10['Team Result Indicator'],
                    cmap = colourbar);

scat5 = ax5.scatter(df_ml_10['Team Av Pass Accuracy Diff'],
                    df_ml_10['Opponent Av Pass Accuracy Diff'],
                    c=df_ml_10['Team Result Indicator'],
```

```python
                    cmap = colourbar);

scat6 = ax6.scatter(df_ml_10['Team Av Goal Diff'],
                    df_ml_10['Opponent Av Goal Diff'],
                    c=df_ml_10['Team Result Indicator'],
                    cmap = colourbar);

#设置所有6个图形的轴和图例
fig.tight_layout(pad=6)

ax1.set(xlabel='Team Average Shots Difference',
        ylabel='Opponent Average Shots');

ax2.set(xlabel='Team Average Shots Inside Box Difference',
        ylabel='Opponent Average Shots Inside Box Difference');

ax3.set(xlabel='Team Average Fouls Difference',
        ylabel='Opponent Average Fouls Difference');

ax4.set(xlabel='Team Average Corners Difference',
        ylabel='Opponent Average Corners Difference');

ax5.set(xlabel='Team Average Pass Accuracy % Difference',
        ylabel='Opponent Average Pass Accuracy % Difference');

ax6.set(xlabel='Team Average Goals Difference',
        ylabel='Opponent Average Goals Difference');

ax_iter = [ax1, ax2, ax3, ax4, ax5, ax6]

for ax in ax_iter:
    ax.legend(*scat2.legend_elements(), title='Target \n Team \n Result', loc='upper right', fontsize='small');
    ax.set_axisbelow(True)
    ax.grid(color='xkcd:light grey')
    lims = [np.min([ax.get_xlim(), ax.get_ylim()]), np.max([ax.get_xlim(), ax.get_ylim()])]
    ax.set_xlim(lims)
    ax.set_ylim(lims)
    #ax.plot(lims, lims, '--', color = '#FFAAAA')
    a_min = lims[0]
    a_max = lims[1]
    mult = lims[1] - lims[0]
    ax.plot([a_min, a_max], [a_min, a_max], '--', color = '#DD7E7E')
    #ax.plot([a_min, a_max], [a_min+0.3*mult, a_max+0.3*mult], '--', color = '#FFCECE')
    #ax.plot([a_min, a_max], [a_min-0.3*mult, a_max-0.3*mult], '--', color = '#FFCECE')

#保存图形
if save_df_10_fig:
    fig.savefig('figures/average_10_games_team_target_result.png')
```

```python
#-------------------------------- FIGURE 2 --------------------------------

#figure 1 - 设置包装器
fig, ((ax1, ax2, ax3),(ax4, ax5, ax6)) = plt.subplots(ncols=3,
                                                      nrows=2,
                                                      figsize=(18,12))
fig.suptitle('Data Averaged Over 5 Games', y=0.99, fontsize=16, fontweight='bold');

#绘制6个数字
scat1 = ax1.scatter(df_ml_5['Team Av Shots Diff'],
                    df_ml_5['Opponent Av Shots Diff'],
                    c=df_ml_5['Team Result Indicator'],
                    cmap = colourbar);

scat2 = ax2.scatter(df_ml_5['Team Av Shots Inside Box Diff'],
                    df_ml_5['Opponent Av Shots Inside Box Diff'],
                    c=df_ml_5['Team Result Indicator'],
                    cmap = colourbar);

scat3 = ax3.scatter(df_ml_5['Team Av Fouls Diff'],
                    df_ml_5['Opponent Av Fouls Diff'],
                    c=df_ml_5['Team Result Indicator'],
                    cmap = colourbar);

scat4 = ax4.scatter(df_ml_5['Team Av Corners Diff'],
                    df_ml_5['Opponent Av Corners Diff'],
                    c=df_ml_5['Team Result Indicator'],
                    cmap = colourbar);

scat5 = ax5.scatter(df_ml_5['Team Av Pass Accuracy Diff'],
                    df_ml_5['Opponent Av Pass Accuracy Diff'],
                    c=df_ml_5['Team Result Indicator'],
                    cmap = colourbar);

scat6 = ax6.scatter(df_ml_5['Team Av Goal Diff'],
                    df_ml_5['Opponent Av Goal Diff'],
                    c=df_ml_5['Team Result Indicator'],
                    cmap = colourbar);

#设置所有6个图形的轴和图例
fig.tight_layout(pad=6)

ax1.set(xlabel='Team Average Shots Difference',
        ylabel='Opponent Average Shots');

ax2.set(xlabel='Team Average Shots Inside Box Difference',
        ylabel='Opponent Average Shots Inside Box Difference');
```

```python
ax3.set(xlabel='Team Average Fouls Difference',
        ylabel='Opponent Average Fouls Difference');

ax4.set(xlabel='Team Average Corners Difference',
        ylabel='Opponent Average Corners Difference');

ax5.set(xlabel='Team Average Pass Accuracy % Difference',
        ylabel='Opponent Average Pass Accuracy % Difference');

ax6.set(xlabel='Team Average Goals Difference',
        ylabel='Opponent Average Goals Difference');

ax_iter = [ax1, ax2, ax3, ax4, ax5, ax6]

for ax in ax_iter:
    ax.legend(*scat2.legend_elements(), title='Target \n Team \n Result', loc='upper right', fontsize='small');
    ax.set_axisbelow(True)
    ax.grid(color='xkcd:light grey')
    lims = [np.min([ax.get_xlim(), ax.get_ylim()]), np.max([ax.get_xlim(), ax.get_ylim()])]
    ax.set_xlim(lims)
    ax.set_ylim(lims)
    #ax.plot(lims, lims, '--', color = '#FFAAAA')
    a_min = lims[0]
    a_max = lims[1]
    mult = lims[1] - lims[0]
    ax.plot([a_min, a_max], [a_min, a_max], '--', color = '#DD7E7E')
    #ax.plot([a_min, a_max], [a_min+0.3*mult, a_max+0.3*mult], '--', color = '#FFCECE')
    #ax.plot([a_min, a_max], [a_min-0.3*mult, a_max-0.3*mult], '--', color = '#FFCECE')

#保存图形
if save_df_5_fig:
    fig.savefig('figures/average_5_games_team_target_result.png')


# -------------------------------- END --------------------------------

print('\n', 'Script runtime:', round(((time.time()-start)/60), 2), 'minutes')
print(' ---------------- END ---------------- \n')
```

执行后会绘制两幅图，分别保存为 average_10_games_team_target_result.png 和 average_5_games_team_target_result.png。其中，在图 average_10_games_team_target_result.png 中绘制了球队 10 场比赛的平均目标成绩，如图 8-3 所示，绿点表示"本球队"获胜，蓝点表示对手获胜。左下象限的圆点表示质量差的球队和对手，左上表示低质量球队和高质量对手，右上表示高质量球队和高质量对手，右下表示高质量球队和低质量对手。通过绘制的可视化图可知，证明所选特征对比赛结果有一些影响，犯规次数似乎与目标结果没有相关性。

average_5_games_team_target_result.png 中绘制了球队 5 场比赛的平均目标成绩，如图 8-4 所示。

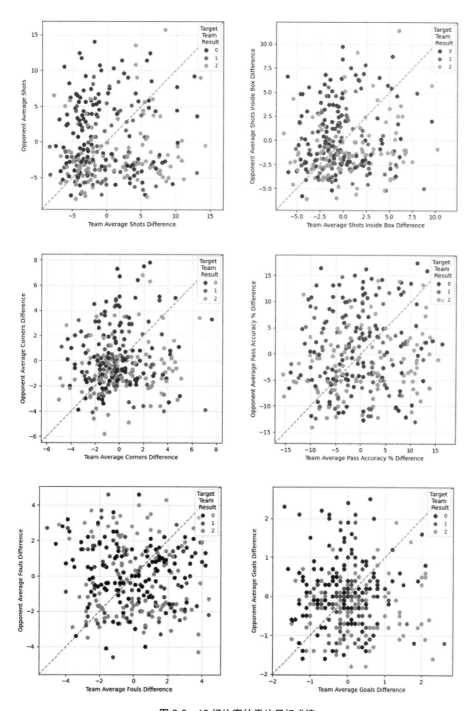

图 8-3　10 场比赛的平均目标成绩

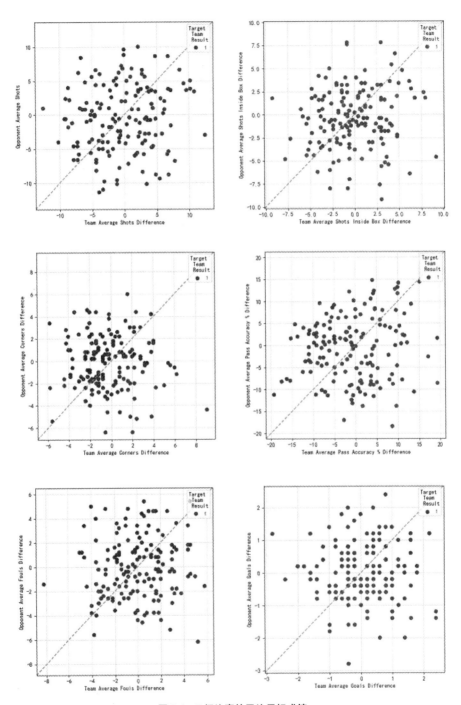

图 8-4　5 场比赛的平均目标成绩

8.5 模型选择和训练

本项目使用 Scikit-Learn 创建机器学习模型并进行训练,所有的模型都使用具有 5 倍交叉验证准确度指标的网格搜索进行优化。

扫码观看本节视频讲解

8.5.1 机器学习函数

在"ml_functions"目录中创建 3 个公用文件,在里面创建训练需要用到的公用函数。
(1) 在实例文件 data_processing.py 中编写如下公用函数。
- 函数 scale_df() 的功能是使用 Scikit-Learn 中的预处理函数重新缩放每个特征,使其平均值和单位向量为零 (mean=0, variance=1),输出是 df 数组,而不是 np 数组。
- 函数 scree_plot(pca_percentages, y_max=40):功能是输入比分规则列表并返回打印参数。

文件 data_processing.py 的具体实现代码如下所示。

```
def scale_df(df, scale, unscale):
    '''
    Parameters
    ----------
    df : pandas DataFrame
        操纵/缩放
    scale : list
        要处理/缩放的索引列表
    unscale : list
        保持不变的列索引引列表
    Returns
    -------
    Scaled df.
    '''
    scaled_np = preprocessing.scale(df)

    col_list = []
    for col in df.columns:
        col_list.append(col)

    scaled_df = pd.DataFrame(scaled_np)
    scaled_df.columns = col_list

    df1 = scaled_df.iloc[:, scale]
    df2 = df.iloc[:, unscale]

    final_df = pd.concat([df1, df2], axis=1, sort=False)

    return final_df
```

```python
def scree_plot(pca_percentages, y_max=40):
    '''
    pca_percentages : list
        主成分百分比变化

    Returns
    -------
    fig : fig
        bar plot.

    '''

    #设置变量
    n_components = len(pca_percentages)

    #实例化图形
    fig, ax = plt.subplots()

    #plot组件
    ax.bar(list(range(1, n_components+1, 1)), pca_percentages, color='paleturquoise',
edgecolor='darkturquoise', zorder=0)

    #用百分数注释
    for p in ax.patches:
        ax.annotate(f'{round(p.get_height(), 1)}%', (p.get_x() + 0.5, p.get_height() + 0.5))

    #绘制各主要部件的线和点
    ax.plot(list(range(1, n_components+1, 1)), pca_percentages, c='firebrick', zorder=1)
    ax.scatter(list(range(1, n_components+1, 1)), pca_percentages, c='firebrick', zorder=2)

    #绘图细节
    fig.suptitle('PCA Scree Plot', y=0.96, fontsize=16, fontweight='bold');
    ax.set(xlabel='Principle Components',
           ylabel='Percentage Variation');
    ax.set_ylim([0,y_max])

    return fig
```

（2）在实例文件 feature_engineering_functions.py 中编写如下公用函数。
- 函数running_mean(x,N)：计算列表x上区间N的滑动平均值。
- 函数average_stats_df()：输出显示平均统计数据的数据帧，包括球队在"比赛次数"上的平均统计数据以及这些比赛中对手的平均统计数据。
- 函数mod_df()：此函数需要函数average_stats_df()的输出内容作为输入，需要输入一个球队和他们的对手（在过去10场比赛中）的平均数据，并删除了只有一个比赛数据的球队。这样做的好处是提供了一个更有用的衡量团队表现的指标。如果"Av Shots Diff"为正值，则表示球队在前几场比赛中的平均投篮次数比对手多，这是机器学习的一个有用特性。

文件 feature_engineering_functions.py 的具体实现代码如下所示。

```python
def running_mean(x, N):
    '''
    Parameters
    ----------
    x : list
        int或float类型的列表
    N : int
        滑动的平均间隔

    Returns
    -------
    list
        平均滑动列表

    '''
    cumsum = np.cumsum(np.insert(x, 0, 0))
    return (cumsum[N:] - cumsum[:-N]) / float(N)

def average_stats_df(games_slide, team_list, team_fixture_id_dict, game_stats, making_predictions=False):
    '''
    Parameters
    ----------
    games_slide : int
        超过平均值的比赛数
    team_list : list
        团队ID的列表.英超联赛应该有20场
    team_fixture_id_dict : dict
        key: 球队id, value: fixture id的列表
    game_stats : nested dict
        key: team id, second-key: fixtue id, value: stats dataframe
    making_predictions: bool
        default = False. 如果创建预测数据帧,则设置为true
    Returns
    -------
    df_ready_for_ml : dataframe
        平均比赛数据
    '''
    if making_predictions:
        x = games_slide
        xx = 1
    else:
        x = -1
        xx = 0
    
    #创建将附加的最终要素
    t_total_shots = []
    t_shots_inside_box = []
    t_fouls = []
```

```python
t_corners = []
t_posession = []
t_pass_accuracy = []
t_goals = []
t_goals_target = []
o_total_shots = []
o_shots_inside_box = []
o_fouls = []
o_corners = []
o_posession = []
o_pass_accuracy = []
o_goals = []
o_goals_target = []
fix_id = []
result_indicator = []
o_team_ID = []

for team_id in team_list[:]:
    team = game_stats[team_id] #球队字典

    #跳过那些比赛较少的球队'games_slide'
    if len(team_fixture_id_dict[team_id]) < games_slide:
        continue

    #创建初始特性,这些特性在每次迭代中都会被覆盖
    team_total_shots = []
    team_shots_inside_box = []
    team_fouls = []
    team_corners = []
    team_posession = []
    team_pass_accuracy = []
    team_goals = []
    opponent_total_shots = []
    opponent_shots_inside_box = []
    opponent_fouls = []
    opponent_corners = []
    opponent_posession = []
    opponent_pass_accuracy = []
    opponent_goals = []
    result_indicator_raw = []

    #迭代fixtureid以创建特征列表
    for game_id in team_fixture_id_dict[team_id]:
        game = team[game_id] #game df
        temp_index = pd.Index(game['Team Identifier'])
        team_ind = temp_index.get_loc(1)
        opponent_ind = temp_index.get_loc(2)
```

```python
#球队和对手伪特征:每场比赛的原始特征数据列表
team_total_shots.append(game['Total Shots'][team_ind])
team_shots_inside_box.append(game['Shots insidebox'][team_ind])
team_fouls.append(game['Fouls'][team_ind])
team_corners.append(game['Corner Kicks'][team_ind])
team_posession.append(game['Ball Possession'][team_ind])
team_pass_accuracy.append(game['Passes %'][team_ind])
team_goals.append(game['Goals'][team_ind])
opponent_total_shots.append(game['Total Shots'][opponent_ind])
opponent_shots_inside_box.append(game['Shots insidebox'][opponent_ind])
opponent_fouls.append(game['Fouls'][opponent_ind])
opponent_corners.append(game['Corner Kicks'][opponent_ind])
opponent_posession.append(game['Ball Possession'][opponent_ind])
opponent_pass_accuracy.append(game['Passes %'][opponent_ind])
opponent_goals.append(game['Goals'][opponent_ind])
result_indicator_raw.append(game['Points'][team_ind])


#使用上面原始特征列表的平均滑动值,以创建最终的特征
team_total_shots_slide = running_mean(team_total_shots, games_slide)[:x]
team_shots_inside_box_slide = running_mean(team_shots_inside_box, games_slide)[:x]
team_fouls_slide = running_mean(team_fouls, games_slide)[:x]
team_corners_slide = running_mean(team_corners, games_slide)[:x]
team_posession_slide = running_mean(team_posession, games_slide)[:x]
team_pass_accuracy_slide = running_mean(team_pass_accuracy, games_slide)[:x]
team_goals_slide = running_mean(team_goals, games_slide)[:x]
team_goals_target = team_goals[games_slide-xx:]
opponent_total_shots_slide = running_mean(opponent_total_shots, games_slide)[:x]
opponent_shots_inside_box_slide = running_mean( opponent_shots_inside_box, games_slide)[:x]
opponent_fouls_slide = running_mean(opponent_fouls, games_slide)[:x]
opponent_corners_slide = running_mean(opponent_corners, games_slide)[:x]
opponent_posession_slide = running_mean(opponent_posession, games_slide)[:x]
opponent_pass_accuracy_slide = running_mean(opponent_pass_accuracy, games_slide)[:x]
opponent_goals_slide = running_mean(opponent_goals, games_slide)[:x]
opponent_goals_target = opponent_goals[games_slide-xx:]
fix_id_slide = team_fixture_id_dict[team_id][games_slide-xx:]
result_indicator_slide = result_indicator_raw[games_slide-xx:]


#在iterables上追加,上述变量将在每次迭代中被覆盖
t_total_shots.extend(team_total_shots_slide)
t_shots_inside_box.extend(team_shots_inside_box_slide)
t_fouls.extend(team_fouls_slide)
t_corners.extend(team_corners_slide)
t_posession.extend(team_posession_slide)
t_pass_accuracy.extend(team_pass_accuracy_slide)
t_goals.extend(team_goals_slide)
t_goals_target.extend(team_goals_target)
o_total_shots.extend(opponent_total_shots_slide)
```

```python
            o_shots_inside_box.extend(opponent_shots_inside_box_slide)
            o_fouls.extend(opponent_fouls_slide)
            o_corners.extend(opponent_corners_slide)
            o_posession.extend(opponent_posession_slide)
            o_pass_accuracy.extend(opponent_pass_accuracy_slide)
            o_goals.extend(opponent_goals_slide)
            o_goals_target.extend(opponent_goals_target)
            fix_id.extend(fix_id_slide)
            result_indicator.extend(result_indicator_slide)
            o_team_ID.append(team_id)

    #将结果拼接到数据帧中
    df_ready_for_ml = pd.DataFrame({})
    df_ready_for_ml['Team Av Shots'] = t_total_shots
    df_ready_for_ml['Team Av Shots Inside Box'] = t_shots_inside_box
    df_ready_for_ml['Team Av Fouls'] = t_fouls
    df_ready_for_ml['Team Av Corners'] = t_corners
    df_ready_for_ml['Team Av Possession'] = t_posession
    df_ready_for_ml['Team Av Pass Accuracy'] = t_pass_accuracy
    df_ready_for_ml['Team Av Goals'] = t_goals
    df_ready_for_ml['Opponent Av Shots'] = o_total_shots
    df_ready_for_ml['Opponent Av Shots Inside Box'] = o_shots_inside_box
    df_ready_for_ml['Opponent Av Fouls'] = o_fouls
    df_ready_for_ml['Opponent Av Corners'] = o_corners
    df_ready_for_ml['Opponent Av Possession'] = o_posession
    df_ready_for_ml['Opponent Av Goals'] = o_goals
    df_ready_for_ml['Opponent Av Pass Accuracy'] = o_pass_accuracy

    df_ready_for_ml['Team Goal Target'] = t_goals_target
    df_ready_for_ml['Opponent Goal Target'] = o_goals_target
    df_ready_for_ml['Target Fixture ID'] = fix_id
    df_ready_for_ml['Result Indicator'] = result_indicator
    if making_predictions:
        df_ready_for_ml['team_id'] = o_team_ID

    #返回数据帧
    return df_ready_for_ml

def mod_df(df, making_predictions=False):
    '''
    Parameters
    ----------
    df : dataframe
        比赛统计,从函数输出: 'average_stats_df()'.
    making_predictions :bool,可选,默认为False.
        默认设置为false,则输出适合于训练模型; 如果设置为true,则输出适合进行预测


    Returns
    -------
```

```
    df_output : dataframe
        修正平均比赛统计数据

    '''

    df_sort = df.sort_values('Target Fixture ID')
    df_sort = df_sort.reset_index(drop=True)

    #在我们的输入数据框(df)中,删除了玩过少于5场或10场比赛的球队的数据
    #然而,我们还没有从已经打了5到10场比赛的oposing球队中删除数据
    #在输入df中,每个赛程有两行,每个队有一行,下面的代码删除了所有只有一个球队统计数据可用的
    比赛的数据,因为这对训练模型没有用处
    #在早期阶段没有这样做,因为这些数据在未来的预测中仍然有用
    if not making_predictions:

        index_to_remove = []

        for i in range(0, len(df_sort)-1):
            if i == 0:
                continue

            elif i == len(df_sort)-1:
                target_m1 = df_sort['Target Fixture ID'].loc[i-1]
                target = df_sort['Target Fixture ID'].loc[i]
                if target != target_m1:
                    index_to_remove.append(i)

            else:
                target_m1 = df_sort['Target Fixture ID'].loc[i-1]
                target = df_sort['Target Fixture ID'].loc[i]
                target_p1 = df_sort['Target Fixture ID'].loc[i+1]
                if (target != target_m1) and (target != target_p1):
                    index_to_remove.append(i)
                else:
                    continue

        df_sort = df_sort.drop(df_sort.index[index_to_remove])


    #创建我们想要的功能
    df_output = pd.DataFrame({})

    df_output['Av Shots Diff'] = df_sort['Team Av Shots'] - df_sort['Opponent Av Shots']
    df_output['Av Shots Inside Box Diff'] = df_sort['Team Av Shots Inside Box'] - df_sort
['Opponent Av Shots Inside Box']
    df_output['Av Fouls Diff'] = df_sort['Team Av Fouls'] - df_sort['Opponent Av Fouls']
    df_output['Av Corners Diff'] = df_sort['Team Av Corners'] - df_sort['Opponent Av
Corners']
    df_output['Av Possession Diff'] = df_sort['Team Av Possession'] - df_sort
```

```python
['Opponent Av Possession']
    df_output['Av Pass Accuracy Diff'] = df_sort['Team Av Pass Accuracy'] - df_sort
['Opponent Av Pass Accuracy']
    df_output['Av Goal Difference'] = df_sort['Team Av Goals'] - df_sort['Opponent Av Goals']
    if not making_predictions:
        df_output['Fixture ID'] = df_sort['Target Fixture ID']
        df_output['Result Indicator'] = df_sort['Result Indicator']
    if making_predictions:
        df_output['Team ID'] = df_sort['team_id']

    return df_output

def combining_fixture_id(df):
    '''
    Parameters
    ----------
    df : dataframe
        game stats, outputted from the funtion: 'mod_df()'.

    Returns
    -------
    df_output : dataframe
        features for both home and away team with target fixture id.

    '''

    #迭代每个对手并添加到上一行
    odd_list = []
    for x in range(1, len(df), 2):
        odd_list.append(x)
    even_list = []
    for x in range(0, len(df)-1, 2):
        even_list.append(x)

    team_df = df.drop(df.index[odd_list])
    team_df = team_df.reset_index(drop=True)
    opponent_df = df.drop(df.index[even_list])
    opponent_df = opponent_df.reset_index(drop=True)

    df_output = pd.DataFrame({})
    df_output['Team Av Shots Diff'] = team_df['Av Shots Diff']
    df_output['Team Av Shots Inside Box Diff'] = team_df['Av Shots Inside Box Diff']
    df_output['Team Av Fouls Diff'] = team_df['Av Fouls Diff']
    df_output['Team Av Corners Diff'] = team_df['Av Corners Diff']
    df_output['Team Av Possession Diff'] = team_df['Av Possession Diff']
    df_output['Team Av Pass Accuracy Diff'] = team_df['Av Pass Accuracy Diff']
    df_output['Team Av Goal Diff'] = team_df['Av Goal Difference']
    df_output['Opponent Av Shots Diff'] = opponent_df['Av Shots Diff']
    df_output['Opponent Av Shots Inside Box Diff'] = opponent_df['Av Shots Inside Box Diff']
```

```
        df_output['Opponent Av Fouls Diff'] = opponent_df['Av Fouls Diff']
        df_output['Opponent Av Corners Diff'] = opponent_df['Av Corners Diff']
        df_output['Opponent Av Possession Diff'] = opponent_df['Av Possession Diff']
        df_output['Opponent Av Pass Accuracy Diff'] = opponent_df['Av Pass Accuracy Diff']
        df_output['Opponent Av Goal Diff'] = opponent_df['Av Goal Difference']
        df_output['Fixture ID'] = team_df['Fixture ID']
        df_output['Team Result Indicator'] = team_df['Result Indicator']
        df_output['Opponent Result Indicator'] = opponent_df['Result Indicator']

        return df_output

def creating_ml_df(df, making_predictions=False):
    '''
    Parameters
    ----------
    df : dataframe
        比赛统计,从函数'average_stats_df()'输出.
    making_predictions: 默认情况下,bool为False.
        默认设置为false,则输出适合于训练模型.如果设置为true,则输出适合进行预测.
    Returns
    -------
    df_output : dataframe
    '''
    modified_df = mod_df(df)
    df_output = combining_fixture_id(modified_df)
    return df_output
```

（3）在实例文件 ml_model_eval.py 中编写如下公用函数。
- 函数 pred_proba_plot()：在给定分类器输出概率的情况下，根据错误给定的结果显示正确预测的结果的直方图。
- 函数 plot_cross_val_confusion_matrix()：根据交叉验证的结果绘制混淆矩阵，与测试分割数据上的标准混淆矩阵相比较。
- 函数 plot_learning_curve()：为指定的估计器绘制学习曲线。

文件 ml_model_eval.py 的具体实现代码如下所示。

```
def pred_proba_plot(clf, X, y, cv=5, no_iter=5, no_bins=25, x_min=0.5, x_max=1, output_progress=True, classifier=''):
    '''
    ----------
    clf : 实现"fit拟合"和"predict预测"的估计器对象,用于拟合数据的对象.

    X : 类似阵列,数据需要进行匹配.例如,可以是列表或至少2d的数组.
    y : 像数组一样,尝试预测的目标变量.
    cv : int,交叉验证生成器或iterable,可选的,确定交叉验证拆分策略,默认值为5.
    no_iter : int,可选的,交叉验证的迭代次数,默认值为5.
    no_bins : int,可选的,直方图的数量,默认值为25.
    x_min : nt,可选的,直方图上的最小x,默认值为0.5.
```

```
        x_max : 英寸,可选的,直方图上的最大x,默认值为1.
        output_progress : 显示进度,可选的,编号迭代完成打印到控制台.默认值为True.
        classifier : 字符串,可选的,分类器被使用后将被输入到标题中.默认值为""(空).

        Returns
        -------
        fig :
        '''

        y_dup = []
        correct_guess_pred = []
        incorrect_guess_pred = []
        for i in range(no_iter):
            if output_progress:
                if i % 2 == 0:
                    print(f'completed {i} iterations')
            skf = StratifiedKFold(n_splits=cv, shuffle=True)
            y_pred_cv = cross_val_predict(clf, X, y, cv=skf)
            y_pred_proba_cv = cross_val_predict(clf, X, y, cv=skf, method='predict_proba')
            y_dup.append(list(y))
            for i in range(len(y_pred_cv)):
                if y_pred_cv[i] == list(y)[i]:
                    correct_guess_pred.append(max(y_pred_proba_cv[i]))
                if y_pred_cv[i] != list(y)[i]:
                    incorrect_guess_pred.append(max(y_pred_proba_cv[i]))
        bins = np.linspace(x_min, x_max, no_bins)
        fig, ax = plt.subplots()
        ax.hist(incorrect_guess_pred, bins, alpha=0.5, edgecolor='#1E212A', color='red',
label='Incorrect Prediction')
        ax.hist(correct_guess_pred, bins, alpha=0.5, edgecolor='#1E212A', color='green',
label='Correct Prediction')
        ax.legend()
        ax.set_title(f'{classifier} - Iterated {no_iter} Times', y=1, fontsize=16,
fontweight='bold');
        ax.set(ylabel='Number of Occurences',
               xlabel='Prediction Probability')
        return fig

    def plot_cross_val_confusion_matrix(clf, X, y, display_labels='', title='', cv=5):
        '''
        根据交叉验证的结果绘制混淆矩阵,与测试分割数据上的标准混淆矩阵相比较.
        Parameters
        ----------
        clf : 实现"fit拟合"和"predict预测"的估计器对象,用于拟合数据的对象.
        X : 类似阵列,数据需要进行匹配.例如,可以是列表或至少2d的数组.
        y : 像数组一样,在监督学习的情况下尝试预测的目标变量.
        display_labels : 形状数组(n_class),可选的,显示用于打印的标签.默认值为""(空).
        title : 字符串,可选的,要显示在绘图顶部的标题,默认值为""(空).
        cv : int,交叉验证生成器或iterable,可选的,确定交叉验证拆分策略,默认值为5.
```

```
    Returns
    -------
    display : :class:'~sklearn.metrics.ConfusionMatrixDisplay'
    '''

    y_pred = cross_val_predict(clf, X, y, cv=cv)
    cm = confusion_matrix(y, y_pred)
    fig = ConfusionMatrixDisplay(confusion_matrix=cm, display_labels=display_labels)
    fig.plot(cmap=plt.cm.Blues)
    fig.ax_.set_title(title)
    return fig

def plot_learning_curve(clf, X, y, scoring='accuracy', training_set_size=5, cv=5, x_min=0,
x_max=500, y_min=0.3, y_max=1.02, title='', leg_loc=4):
    '''
    为给定估计器绘制学习曲线

    Parameters
    ----------
    clf : 实现"fit拟合"和"predict预测"的估计器对象,用于拟合数据的对象.
    X : 类似阵列,数据需要进行匹配.例如,可以是列表或至少2d的数组.
    y : 像数组一样,尝试预测的目标变量.
    scoring : 选择的标准,指标可以是"准确性""损失"等.必须是字符串!，默认值为"accuracy".
    training_set_size : 整数,可选的,训练数组上进行的拆分次数.默认值为5.
    cv : 交叉验证生成器或iterable,可选的,确定交叉验证拆分策略.默认值为5.
    x_min : int,可选的,图上的最小x.默认值为0.
    x_max : int, 可选的,图上的最大x.默认值为500.
    y_min : int,可选的,图上的y-min.默认值为0.3.
    y_max : int, 可选的,图上的y-max.默认值为1.02.
    title : string, 可选的,图的标题.默认值为""（空）.
    leg_loc : int, 可选的,图例在图中的位置.默认值为4.

    Returns
    -------
    fig

    '''

    #使用50种不同的训练集计算训练和val精度
    train_size,train_scores,valid_scores = learning_curve(clf, X, y, cv=cv, random_state=42,
train_sizes=np.linspace(0.1, 1.0, training_set_size), scoring=scoring)

    #计算训练集和val集的平均值和标准差
    train_mean = np.mean(train_scores,axis=1)
    valid_mean = np.mean(valid_scores,axis=1)
    valid_std = np.std(valid_scores,axis=1)

    #绘制线
```

```python
fig, ax = plt.subplots()
ax.plot(train_size, train_mean, color="royalblue", label="Training score")
ax.plot(train_size, valid_mean, '--', color="#111111", label="Cross-validation score")

#绘制误差带
ax.fill_between(train_size, valid_mean - valid_std, valid_mean + valid_std, color="#DDDDDD")

#绘图细节
ax.set_xlabel("Training Set Size")
ax.set_ylabel("Accuracy Score")
ax.legend(loc=leg_loc)

#组织轴
ax.set_title(title, y=1, fontsize=14, fontweight='bold');
ax.set_ylim(y_min,y_max)
ax.set_xlim(x_min,x_max)

return fig
```

8.5.2 数据降维

编写实例文件 dimensionality_reduction.py，功能是实现数据降维。数据降维直观的好处是维度降低了，便于计算和可视化，其更深层次的意义在于有效信息的综合提取及无用信息的摒弃。文件 dimensionality_reduction.py 的具体实现代码如下所示。

```python
df_10_saved_name = '2019_2020_prem_df_for_ml_10_v2.txt'

plot_scree_plot = True
save_scree_plot = False

xplot_pc1_pc2 = True
save_pc1_pc2 = False

xplot_ld1_ld2 = True
save_ld1_ld2 = False

#------------------------- PRINCIPLE COMPONENT ANALYSIS -------------------------

with open(f'../prem_clean_fixtures_and_dataframes/{df_10_saved_name}', 'rb') as myFile:
    df_ml_10 = pickle.load(myFile)

#缩放数据帧,使所有特征具有零均值和单位向量。这在PCA之前是必要的,因为使用了欧几里得距离
df_ml_10 = scale_df(df_ml_10, list(range(14)), [14,15,16])

#创建目标和feature df
x_10 = df_ml_10.drop(['Fixture ID', 'Team Result Indicator', 'Opponent Result Indicator'], axis=1)
```

```python
    y_10 = df_ml_10['Team Result Indicator']

#打印每个主要组成部分的贡献/变化
n_components = 6
pca = PCA(n_components=n_components)
pca.fit(x_10)
pca_percentages = list(pca.explained_variance_ratio_)
pca_percentages = [element * 100 for element in pca_percentages]
for i in range(0, n_components, 1):
    print(f'PCA{i+1}:', round(pca_percentages[i], 2), '%')

if plot_scree_plot:
    fig = scree_plot(pca_percentages)
    if save_scree_plot:
        fig.savefig('figures/PCA_Scree_Plot_ml10.png')

# ---------- X-plot PC1 and PC2 ----------

#创建变量
pca_values = pca.fit_transform(x_10)

if xplot_pc1_pc2:

    #实例化图形并绘制散点图
    fig, ax = plt.subplots()
    scat = ax.scatter(pca_values[:,0],
                      pca_values[:,1],
                      c=df_ml_10['Team Result Indicator'],
                      cmap='winter');

    #fig详细信息
    fig.suptitle('PCA X-Plot', y=0.96, fontsize=16, fontweight='bold');
    ax.set(xlabel='PC1', ylabel='PC2');
    ax.legend(*scat.legend_elements(),
              title='Target \n Team \n Result',
              loc='upper right',
              fontsize='small')
    ax.grid(color='xkcd:light grey')
    ax.set_axisbelow(True)

    if save_pc1_pc2:
        fig.savefig('figures/PC1_PC2_xplot_ml10.png')

#----------------------- LINEAR DISCRIMINANT ANALYSIS ------------------------

#实例化和拟合LDA分类器,以达到降维的目的
clf = LinearDiscriminantAnalysis(n_components=2)
clf.fit(x_10, y_10)
```

```python
# ---------- X-plot LDA1 and LDA2 ----------

#创建变量
lda_values = clf.fit_transform(x_10, y_10)

if xplot_ld1_ld2:

    #实例化图形并绘制散点图
    fig, ax = plt.subplots()
    scat = ax.scatter(lda_values[:,0],
                      lda_values[:,1],
                      c=df_ml_10['Team Result Indicator'],
                      cmap='winter');

    #fig详情
    fig.suptitle('LDA X-Plot', y=0.96, fontsize=16, fontweight='bold');
    ax.set(xlabel='LD1', ylabel='LD2');
    ax.legend(*scat.legend_elements(),
              title='Target \n Team \n Result',
              loc='upper right',
              fontsize='small')

    ax.grid(color='xkcd:light grey')
    ax.set_axisbelow(True)

    if save_ld1_ld2:
        fig.savefig('figures/LDA_xplot_ml10.png')
#---------------------------------- END -------------------------------------

print('\n', 'Script runtime:', round(((time.time()-start)/60), 2), 'minutes')
print(' ---------------- END ---------------- \n')
```

执行后分别绘制 LDA X-Plot、PCA Scree Plot 和 PCA X-Plot 图，如图 8-5 所示。

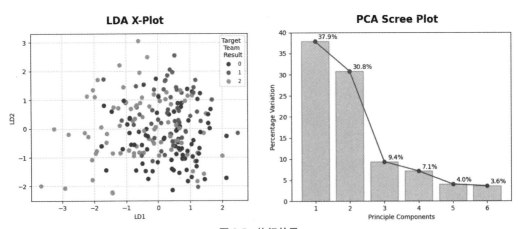

图 8-5　执行效果

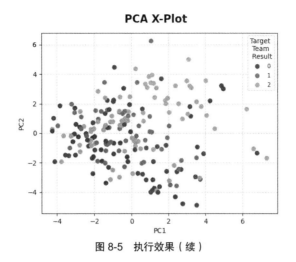

图 8-5 执行效果（续）

8.5.3 MLP 神经网络

编写实例文件 MLP_sklearn.py，功能是使用 MLP 神经网络训练数据和进行评估，并绘制出可视化图形。文件 MLP_sklearn.py 的具体实现代码如下所示。

```
df_5_saved_name = '2019_2020_prem_df_for_ml_5_v2.txt'
df_10_saved_name = '2019_2020_prem_df_for_ml_10_v2.txt'

grid_search = False
save_grid_search_fig = False

create_final_model = True

#------------------------------ ML MODEL BUILD ------------------------------

#导入数据并创建要素数据框和目标系列

with open(f'../prem_clean_fixtures_and_dataframes/{df_5_saved_name}', 'rb') as myFile:
    df_ml_5 = pickle.load(myFile)

with open(f'../prem_clean_fixtures_and_dataframes/{df_10_saved_name}', 'rb') as myFile:
    df_ml_10 = pickle.load(myFile)

#缩放数据帧以使所有特征具有零平均值和单位向量
df_ml_10 = scale_df(df_ml_10, list(range(14)), [14,15,16])
df_ml_5 = scale_df(df_ml_5, list(range(14)), [14,15,16])

#为df_10创建Fixtures和标签df
```

```python
    x_10 = df_ml_10.drop(['Fixture ID', 'Team Result Indicator', 'Opponent Result Indicator'],
axis=1)
    y_10 = df_ml_10['Team Result Indicator']

    #为df_5创建Fixtures和标签df
    x_5 = df_ml_5.drop(['Fixture ID', 'Team Result Indicator', 'Opponent Result Indicator'],
axis=1)
    y_5 = df_ml_5['Team Result Indicator']


    #--------------------------- MULTILAYER PERCEPTRON ---------------------------

    #分为训练数据和测试数据
    x_train, x_test, y_train, y_test = train_test_split(x_10, y_10, test_size=0.2)

    #实例化MLP分类器,并使其适合数据
    clf = MLPClassifier(hidden_layer_sizes=(18, 12),
                        activation='logistic',
                        random_state=0,
                        max_iter=5000)
    clf.fit(x_train, y_train)

    #打印交叉验证准确性得分
    cv_score_av = round(np.mean(cross_val_score(clf, x_10, y_10))*100,1)
    print('Cross-Validation Accuracy Score ML10: ', cv_score_av, '%\n')


    #-------------------------------- GRID SEARCH --------------------------------

    if grid_search:
        #创建元组列表以测试隐藏层的长度
        hidden_layer_test = []
        for i in range(6,20,2):
            a = list(range(6,30,4))
            b = [i] * len(a)
            c = list(zip(a, b))
            hidden_layer_test.extend(c)

        param_grid_grad = [{'hidden_layer_sizes':hidden_layer_test}]

        #MLP网格搜索
        grid_search_grad = GridSearchCV(clf,
                                        param_grid_grad,
                                        cv=5,
                                        scoring = 'accuracy',
                                        return_train_score = True)
        grid_search_grad.fit(x_10, y_10)

        #从网格搜索中输出最佳交叉验证分数和参数
```

```python
    print('\n', 'Gradient Best Params: ' , grid_search_grad.best_params_)
    print('Gradient Best Score: ' , grid_search_grad.best_score_ , '\n')

    print(grid_search_grad.cv_results_['mean_test_score'])


# ---------- PLOTTING THE DATA ----------

if grid_search:
    #用x、y、z数据填充df
    matrix_plot_data = pd.DataFrame({})
    matrix_plot_data['x'] = list(zip(*hidden_layer_test))[0]
    matrix_plot_data['y'] = list(zip(*hidden_layer_test))[1]
    matrix_plot_data['z'] = grid_search_grad.cv_results_['mean_test_score']

    #将z列表转换为矩阵格式
    Z = matrix_plot_data.pivot_table(index='x', columns='y', values='z').T.values

    #获取x轴和y轴
    X_unique = np.sort(matrix_plot_data.x.unique())
    Y_unique = np.sort(matrix_plot_data.y.unique())

    #使用seaborn实例化图形并绘制热图
    fig, ax = plt.subplots()
    im = sns.heatmap(Z, annot=True,linewidths=.5)

    #标记x和y以及标题
    ax.set_xticklabels(X_unique)
    ax.set_yticklabels(Y_unique)
    ax.set(xlabel='Hidden Layer 1 Length',
           ylabel='Hidden Layer 2 Length');
    fig.suptitle('Cross Val Accuracy', y=0.95, fontsize=16, fontweight='bold');

    if save_grid_search_fig:
        fig.savefig('figures/testing_hidden_layer_lengths.png')



#-------------------------------- FINAL MODEL --------------------------------

#将从上述超参数测试中学习,并使用100%的数据训练最终模型。该模型可用于未来的预测
if create_final_model:

    #维护和训练df_5网络
    ml_5_mlp = MLPClassifier(hidden_layer_sizes=(18, 12),
                             activation='logistic',
                             random_state=0,
                             max_iter=5000)
    ml_5_mlp.fit(x_5, y_5)
```

```
#维护和训练df_10网络
ml_10_mlp = MLPClassifier(hidden_layer_sizes=(18, 12),
                          activation='logistic',
                          random_state=0,
                          max_iter=5000)
ml_10_mlp.fit(x_10, y_10)

with open('ml_models/mlp_model_5.pk1', 'wb') as myFile:
    pickle.dump(ml_5_mlp, myFile)

with open('ml_models/mlp_model_10.pk1', 'wb') as myFile:
    pickle.dump(ml_10_mlp, myFile)

# -------------------------------- END --------------------------------

print('\n', 'Script runtime:', round(((time.time()-start)/60), 2), 'minutes')
print(' ---------------- END ---------------- \n')
```

执行后绘制可视化结果,并保存为 testing_hidden_layer_lengths.png,如图 8-6 所示。

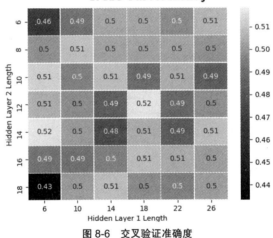

图 8-6 交叉验证准确度

8.6 模型评估

除了准确性之外,本项目绘制了 5 倍交叉验证结果的混淆矩阵来评估模型性能。在本实例中,分别使用 3 种方法实现了模型评估:随机森林模型、SVM 模型和近邻模型,并分别统计了准确度。

扫码观看本节视频讲解

8.6.1 近邻模型和混淆矩阵模型

编写实例文件 nearest_neighbor_model_build.py，功能是实现近邻模型和混淆矩阵模型的评估。文件 nearest_neighbor_model_build.py 的具体实现流程如下所示。

（1）准备数据集，然后缩放数据帧，使所有特征具有零均值和单位向量。这一点很重要，因为算法中使用了欧几里得距离，因此我们希望对每个特征平均加权。代码如下：

```
df_5_saved_name = '2019_2020_prem_df_for_ml_5_v2.txt'
df_10_saved_name = '2019_2020_prem_df_for_ml_10_v2.txt'

test_n_neighbors = False

pred_prob_plot_df10 = False
save_pred_prob_plot_df10 = False
pred_prob_plot_df5 = False
save_pred_prob_plot_df5 = False

save_conf_matrix_df10 = False
save_conf_matrix_df5 = False

save_learning_curve_df10 = False
save_learning_curve_df5 = False

create_final_model = True

#------------------------------ ML MODEL BUILD --------------------------------

#导入数据并创建数据框和目标系列元素

with open(f'../prem_clean_fixtures_and_dataframes/{df_5_saved_name}', 'rb') as myFile:
    df_ml_5 = pickle.load(myFile)

with open(f'../prem_clean_fixtures_and_dataframes/{df_10_saved_name}', 'rb') as myFile:
    df_ml_10 = pickle.load(myFile)

#缩放数据帧,使所有特征具有零均值和单位向量
df_ml_10 = scale_df(df_ml_10, list(range(14)), [14,15,16])
df_ml_5 = scale_df(df_ml_5, list(range(14)), [14,15,16])

x_10 = df_ml_10.drop(['Fixture ID', 'Team Result Indicator', 'Opponent Result Indicator'], axis=1)
y_10 = df_ml_10['Team Result Indicator']

x_5 = df_ml_5.drop(['Fixture ID', 'Team Result Indicator', 'Opponent Result Indicator'], axis=1)
y_5 = df_ml_5['Team Result Indicator']
```

（2）开始实现 K 近邻算法，分别使用不同的输入数据来训练模型和测试模型。代码如下：

```python
def k_nearest_neighbor_train(df, print_result=True, print_result_label=''):

    #创建特征矩阵
    x = df.drop(['Fixture ID', 'Team Result Indicator', 'Opponent Result Indicator'], axis=1)
    y = df['Team Result Indicator']

    #分为训练数据和测试数据
    x_train, x_test, y_train, y_test = train_test_split(x, y, test_size=0.2)

    #实例化K最近邻类
    clf = KNeighborsClassifier(n_neighbors=11, weights='distance')

    #训练模型
    clf.fit(x_train, y_train)

    if print_result:
        print(print_result_label)
        #训练数据
        train_data_score = round(clf.score(x_train, y_train) * 100, 1)
        print(f'Training data score = {train_data_score}%')

        #测试数据
        test_data_score = round(clf.score(x_test, y_test) * 100, 1)
        print(f'Test data score = {test_data_score}% \n')

    return clf, x_train, x_test, y_train, y_test


ml_10_knn, x10_train, x10_test, y10_train, y10_test = k_nearest_neighbor_train(df_ml_10, print_result_label='DF_ML_10')
ml_5_knn, x5_train, x5_test, y5_train, y5_test = k_nearest_neighbor_train(df_ml_5, print_result_label='DF_ML_5')


# ---------- ENSEMBLE MODELLING ----------

#将结合使用相同算法的结果,但使用不同的输入数据来训练模型
#这些特征大体上仍然相同,但在不同的比赛数中平均值为df_ml_10是10个比赛,df_ml_5是5个比赛
#减少df_ml_5中的Fixture,使其仅包含df_ml_10中的Fixture,并训练新数据集
df_ml_5_dropto10 = df_ml_5.drop(list(range(0,50)))
ml_5_to10_knn, x5_to10_train, x5_to10_test, y5_to10_train, y5_to10_test = k_nearest_neighbor_train(df_ml_5_dropto10, print_result=False)

#独立使用两个df输入进行预测
y_pred_ml10 = ml_10_knn.predict(x10_test)
y_pred_ml5to10 = ml_5_to10_knn.predict(x10_test)
```

```python
#独立地对每个数据集进行概率预测
pred_proba_ml10 = ml_10_knn.predict_proba(x10_test)
pred_proba_ml5_10 = ml_5_to10_knn.predict_proba(x10_test)

#组合独立概率和创建组合类预测
pred_proba_ml5and10 = (np.array(pred_proba_ml10) + np.array(pred_proba_ml5_10)) / 2.0
y_pred_ml5and10 = np.argmax(pred_proba_ml5and10, axis=1)

#准确度得分变量
y_pred_ml10_accuracy = round(accuracy_score(y10_test, y_pred_ml10), 3) * 100
y_pred_ml5to10_accuracy = round(accuracy_score(y10_test, y_pred_ml5to10), 3) * 100
y_pred_ml5and10_accuracy = round(accuracy_score(y10_test, y_pred_ml5and10), 3) * 100

print('ENSEMBLE MODEL TESTING')
print(f'Accuracy of df_10 alone = {y_pred_ml10_accuracy}%')
print(confusion_matrix(y10_test, y_pred_ml10), '\n')
print(f'Accuracy of df_5 alone = {y_pred_ml5to10_accuracy}%')
print(confusion_matrix(y10_test, y_pred_ml5to10), '\n')
print(f'Accuracy of df_5 and df_10 combined = {y_pred_ml5and10_accuracy}%')
print(confusion_matrix(y10_test, y_pred_ml5and10), '\n\n')
```

(3)基于 ml_10 测试 k 均值,并绘制可视化结果图。代码如下:

```python
#下面的代码在内环中的n个邻居上迭代,在外环中的i个不同的训练测试拆分。增加外环的范围以获得更稳定的结果
if test_n_neighbors:
    
    test_accuracy_compiled = []
    for i in range(1, 10, 1):
        test_accuracy = []
        for n in range(1, 50, 1):
            x_train, x_test, y_train, y_test = train_test_split(x_10, y_10, test_size=0.2)
            clf = KNeighborsClassifier(n_neighbors=n, weights='uniform')
            clf.fit(x_train, y_train)
            test_accuracy.append(round(clf.score(x_test, y_test) * 100, 1))
        test_accuracy_compiled.append(test_accuracy)
    test_accuracy_compiled_np = np.transpose(np.array(test_accuracy_compiled))
    test_accuracy_compiled_av = np.mean(test_accuracy_compiled_np, axis=1)
    
    fig, ax = plt.subplots()
    ax.plot(range(1,50, 1), test_accuracy_compiled_av, label='Weights = Uniform')
    ax.set_xlabel('n_neighbors')
    ax.set_ylabel('Accuracy Score %')
    ax.set_title('Testing k values ml_10', y=1, fontsize=14, fontweight='bold');
    ax.legend(loc=4)
    plt.savefig('figures\ml_10_testing_k_values_uniform.png')
```

执行后会创建文件 ml_10_testing_k_values_uniform.png,如图 8-7 所示。

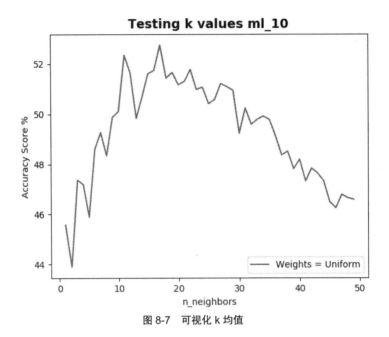

图 8-7 可视化 k 均值

（4）基于 ml_10 绘制可视化最近邻混淆矩阵，代码如下：

```
#交叉验证
skf = StratifiedKFold(n_splits=5, shuffle=True)

cv_score_av = round(np.mean(cross_val_score(ml_10_knn, x_10, y_10, cv=skf))*100,1)
print('Cross-Validation Accuracy Score ML10: ', cv_score_av, '%\n')

cv_score_av = round(np.mean(cross_val_score(ml_5_knn, x_5, y_5, cv=skf))*100,1)
print('Cross-Validation Accuracy Score ML5: ', cv_score_av, '%\n')


# ---------- PREDICTION PROBABILITY PLOTS ----------

if pred_prob_plot_df10:
    fig = pred_proba_plot(ml_10_knn,
                          x_10,
                          y_10,
                          no_iter=5,
                          no_bins=36,
                          x_min=0.3,
                          classifier='Nearest Neighbor (ml_10)')
    if save_pred_prob_plot_df10:
        fig.savefig('figures/ml_10_nearest_neighbor_pred_proba.png')
```

执行后会创建文件 ml_10_nearest_neighbor_pred_proba.png，如图 8-8 所示。

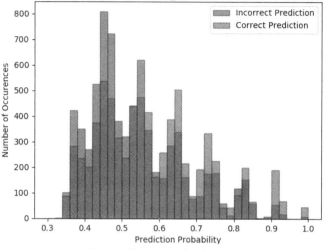

图 8-8　基于 ml-10 的近邻混淆矩阵

（5）基于 ml-5 绘制可视化最近邻混淆矩阵，代码如下：

```
if pred_prob_plot_df5:
    fig = pred_proba_plot(ml_5_knn,
                          x_5,
                          y_5,
                          no_iter=50,
                          no_bins=36,
                          x_min=0.3,
                          classifier='Nearest Neighbor (ml_5)')
    if save_pred_prob_plot_df10:
        fig.savefig('figures/ml_5_nearest_neighbor_pred_proba.png')


# ---------- CONFUSION MATRIX PLOTS ----------

plot_cross_val_confusion_matrix(ml_10_knn,
                                x_10,
                                y_10,
                                display_labels=('team loses', 'draw', 'team wins'),
                                title='Nearest Neighbor Confusion Matrix ML10',
                                cv=skf)
if save_conf_matrix_df10:
    plt.savefig('figures\ml_10_confusion_matrix_cross_val_nearest_neighbor.png')

plot_cross_val_confusion_matrix(ml_5_knn,
                                x_5,
                                y_5,
```

```
                            display_labels=('team loses', 'draw', 'team wins'),
                            title='Nearest Neighbor Confusion Matrix ML5',
                            cv=skf)
if save_conf_matrix_df5:
    plt.savefig('figures\ml_5_confusion_matrix_cross_val_nearest_neighbor.png')
```

（6）基于 ml_10 绘制可视化曲线图，代码如下：

```
plot_learning_curve(ml_10_knn,
                    x_10,
                    y_10,
                    training_set_size=10,
                    x_max=240,
                    title='Learning Curve - Nearest Neighbor DF_10',
                    leg_loc=1)
if save_learning_curve_df10:
    plt.savefig('figures\ml_10_nearest_neighbor_learning_curve.png')
```

执行效果如图 8-9 所示。

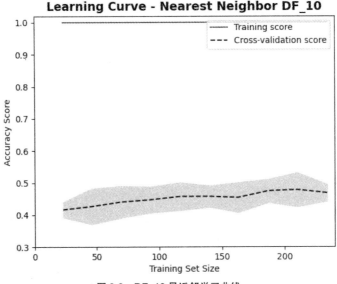

图 8-9　DF_10 最近邻学习曲线

（7）基于 ml_5 绘制可视化曲线图，代码如下：

```
plot_learning_curve(ml_5_knn,
                    x_5,
                    y_5,
                    training_set_size=10,
                    x_max=280,
                    title='Learning Curve - Nearest Neighbor DF_5',
```

```
        leg_loc=1)
if save_learning_curve_df5:
    plt.savefig('figures\ml_5_nearest_neighbor_learning_curve.png')
```

执行效果如图 8-10 所示。

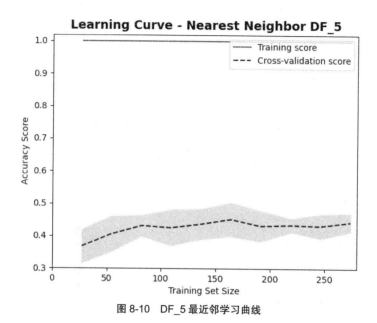

图 8-10　DF_5 最近邻学习曲线

（8）生成模型，从上述超参数进行测试学习，并使用 100% 的数据训练最终模型，该模型可用于未来的预测。代码如下：

```
if create_final_model:
    
    #维护和训练df_5网络
    ml_5_knn = KNeighborsClassifier(n_neighbors=11, weights='distance')
    ml_5_knn.fit(x_5, y_5)
    
    #维护和训练df_10网络
    ml_10_knn = KNeighborsClassifier(n_neighbors=11, weights='distance')
    ml_10_knn.fit(x_10, y_10)
    
    with open('ml_models/knn_model_5.pk1', 'wb') as myFile:
        pickle.dump(ml_5_knn, myFile)
    
    with open('ml_models/knn_model_10.pk1', 'wb') as myFile:
        pickle.dump(ml_10_knn, myFile)

# -------------------------------- END --------------------------------
```

```
print('\n', 'Script runtime:', round(((time.time()-start)/60), 2), 'minutes')
print(' ----------------- END ----------------- \n')
```

8.6.2 随机森林模型和混淆矩阵模型

编写实例文件 random_forest_model_build.py,功能是实现随机森林模型和混淆矩阵模型的评估。文件 random_forest_model_build.py 的具体实现流程如下所示。

(1) 缩放数据帧,使所有特征具有零均值和单位向量。代码如下:

```
df_ml_10 = scale_df(df_ml_10, list(range(14)), [14,15,16])
df_ml_5 = scale_df(df_ml_5, list(range(14)), [14,15,16])

x_10 = df_ml_10.drop(['Fixture ID', 'Team Result Indicator', 'Opponent Result Indicator'], axis=1)
y_10 = df_ml_10['Team Result Indicator']

x_5 = df_ml_5.drop(['Fixture ID', 'Team Result Indicator', 'Opponent Result Indicator'], axis=1)
y_5 = df_ml_5['Team Result Indicator']
```

(2) 编写随机森林函数,创建 features 矩阵,然后训练模型并测试数据。代码如下:

```
def rand_forest_train(df, print_result=True, print_result_label=''):

    #创建features矩阵
    x = df.drop(['Fixture ID', 'Team Result Indicator', 'Opponent Result Indicator'], axis=1)
    y = df['Team Result Indicator']

    #实例化随机森林类
    clf = RandomForestClassifier(max_depth=4, max_features=4, n_estimators=120)

    #分为训练数据和测试数据
    x_train, x_test, y_train, y_test = train_test_split(x, y, test_size=0.2)

    #训练模型
    clf.fit(x_train, y_train)

    if print_result:
        print(print_result_label)
        #训练数据
        train_data_score = round(clf.score(x_train, y_train) * 100, 1)
        print(f'Training data score = {train_data_score}%')

        #测试数据
        test_data_score = round(clf.score(x_test, y_test) * 100, 1)
        print(f'Test data score = {test_data_score}% \n')
```

```
    return clf, x_train, x_test, y_train, y_test

ml_10_rand_forest, x10_train, x10_test, y10_train, y10_test = rand_forest_train(df_ml_10, print_result_label='DF_ML_10')
ml_5_rand_forest, x5_train, x5_test, y5_train, y5_test = rand_forest_train(df_ml_5, print_result_label='DF_ML_5')
```

(3)集成建模,使用相同的随机森林算法、不同的输入数据来训练模型。这些特征大体上仍然相同,但在不同的比赛数中平均值为 df_ml_10 表示 10 场比赛,df_ml_5 表示 5 场比赛。代码如下:

```
#减少df_ml_5中的fixtures,使其仅包含df_ml_10中的features,并训练新数据集
df_ml_5_dropto10 = df_ml_5.drop(list(range(0,50)))
ml_5_to10_rand_forest, x5_to10_train, x5_to10_test, y5_to10_train, y5_to10_test = rand_forest_train(df_ml_5_dropto10, print_result=False)

#独立使用两个df输入进行预测
y_pred_ml10 = ml_10_rand_forest.predict(x10_test)
y_pred_ml5to10 = ml_5_to10_rand_forest.predict(x10_test)

#独立地对每个数据集进行概率预测
pred_proba_ml10 = ml_10_rand_forest.predict_proba(x10_test)
pred_proba_ml5_10 = ml_5_to10_rand_forest.predict_proba(x10_test)

#组合独立概率和创建组合类预测
pred_proba_ml5and10 = (np.array(pred_proba_ml10) + np.array(pred_proba_ml5_10)) / 2.0
y_pred_ml5and10 = np.argmax(pred_proba_ml5and10, axis=1)

#准确度得分变量
y_pred_ml10_accuracy = round(accuracy_score(y10_test, y_pred_ml10), 3) * 100
y_pred_ml5to10_accuracy = round(accuracy_score(y10_test, y_pred_ml5to10), 3) * 100
y_pred_ml5and10_accuracy = round(accuracy_score(y10_test, y_pred_ml5and10), 3) * 100

print('ENSEMBLE MODEL TESTING')
print(f'Accuracy of df_10 alone = {y_pred_ml10_accuracy}%')
print(confusion_matrix(y10_test, y_pred_ml10), '\n')
print(f'Accuracy of df_5 alone = {y_pred_ml5to10_accuracy}%')
print(confusion_matrix(y10_test, y_pred_ml5to10), '\n')
print(f'Accuracy of df_5 and df_10 combined = {y_pred_ml5and10_accuracy}%')
print(confusion_matrix(y10_test, y_pred_ml5and10), '\n\n')
```

(4)实现网格搜索功能,代码如下:

```
if grid_search:
    param_grid_grad = [{'n_estimators':list(range(50,200,50)),
                        'max_depth':list(range(1,5,1)),
                        'max_features':list(range(2,5,1))}]
    param_grid_grad = [{'n_estimators':list(range(10,200,10))}]
```

```python
#随机森林网格搜索
grid_search_grad = GridSearchCV(ml_10_rand_forest,
                                param_grid_grad,
                                cv=5,
                                scoring = 'accuracy',
                                return_train_score = True)
grid_search_grad.fit(x_10, y_10)

#从网格搜索中输出最佳交叉验证分数和参数
print('\n', 'Gradient Best Params: ' , grid_search_grad.best_params_)
print('Gradient Best Score: ' , grid_search_grad.best_score_ , '\n')
```

（5）实现模型评估和交叉验证功能，然后绘制不同的可视化预测评估图。代码如下：

```python
#交叉验证
skf = StratifiedKFold(n_splits=5, shuffle=True)

cv_score_av = round(np.mean(cross_val_score(ml_10_rand_forest, x_10, y_10, cv=skf))*100,1)
print('Cross-Validation Accuracy Score ML10: ', cv_score_av, '%\n')

cv_score_av = round(np.mean(cross_val_score(ml_5_rand_forest, x_5, y_5, cv=skf))*100,1)
print('Cross-Validation Accuracy Score ML5: ', cv_score_av, '%\n')


# ---------- 预测概率图----------

if pred_prob_plot_df10:
    fig = pred_proba_plot(ml_10_rand_forest,
                          x_10,
                          y_10,
                          no_iter=50,
                          no_bins=36,
                          x_min=0.3,
                          classifier='Random Forest (ml_10)')
    if save_pred_prob_plot_df10:
        fig.savefig('figures/ml_10_random_forest_pred_proba.png')

if pred_prob_plot_df5:
    fig = pred_proba_plot(ml_5_rand_forest,
                          x_5,
                          y_5,
                          no_iter=50,
                          no_bins=35,
                          x_min=0.3,
                          classifier='Random Forest (ml_5)')
    if save_pred_prob_plot_df5:
        fig.savefig('figures/ml_5_random_forest_pred_proba.png')
```

```python
# ---------- CONFUSION MATRIX PLOTS ----------

#获取交叉验证结果

plot_cross_val_confusion_matrix(ml_10_rand_forest,
                                x_10,
                                y_10,
                                display_labels=('team loses', 'draw', 'team wins'),
                                title='Random Forest Confusion Matrix ML10',
                                cv=skf)
if save_conf_matrix_df10:
    plt.savefig('figures\ml_10_confusion_matrix_cross_val_random_forest.png')

plot_cross_val_confusion_matrix(ml_5_rand_forest,
                                x_5,
                                y_5,
                                display_labels=('team loses', 'draw', 'team wins'),
                                title='Random Forest Confusion Matrix ML5',
                                cv=skf)
if save_conf_matrix_df5:
    plt.savefig('figures\ml_5_confusion_matrix_cross_val_random_forest.png')


# ---------- 学习曲线图 ----------

plot_learning_curve(ml_10_rand_forest,
                    x_10,
                    y_10,
                    training_set_size=20,
                    x_max=240,
                    title='Learning Curve - Random Forest DF_10')
if save_learning_curve_df10:
    plt.savefig('figures\ml_10_random_forest_learning_curve.png')

plot_learning_curve(ml_5_rand_forest,
                    x_5,
                    y_5,
                    training_set_size=20,
                    x_max=280,
                    title='Learning Curve - Random Forest DF_5')
if save_learning_curve_df5:
    plt.savefig('figures\ml_5_random_forest_learning_curve.png')


# ---------- 重要性FEATURE ----------

fi_ml_10 = pd.DataFrame({'feature': list(x10_train.columns),'importance': ml_10_rand_forest.feature_importances_}).sort_values('importance', ascending = False)
```

```
    fi_ml_5 = pd.DataFrame({'feature': list(x5_train.columns),'importance': ml_5_rand_forest.
feature_importances_}).sort_values('importance', ascending = False)
```

执行后会绘制不同的可视化评估图，例如 DF_5 随机森林学习曲线如图 8-11 所示。

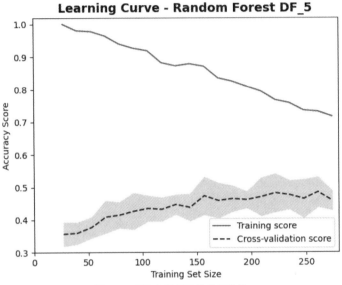

图 8-11 DF_5 随机森林学习曲线

8.6.3 SVM 模型和混淆矩阵模型

编写实例文件 support_vector_machine_model_build.py，功能是实现 SVM 模型和混淆矩阵模型的评估。文件 support_vector_machine_model_build.py 的具体实现流程如下所示。

（1）创建需要的多个输入变量，包括数据文件名、保存可视化图参数等。代码如下：

```
df_5_saved_name = '2019_2020_prem_df_for_ml_5_v2.txt'
df_10_saved_name = '2019_2020_prem_df_for_ml_10_v2.txt'

pred_prob_plot_df10 = False
save_pred_prob_plot_df10 = False
pred_prob_plot_df5 = False
save_pred_prob_plot_df5 = False

save_conf_matrix_df10 = False
save_conf_matrix_df5 = False

save_learning_curve_df10 = False
```

```
save_learning_curve_df5 = False

create_final_model = True
```

（2）构建 ML 模型，导入数据并创建要素数据框和目标。代码如下：

```
with open(f'../prem_clean_fixtures_and_dataframes/{df_5_saved_name}', 'rb') as myFile:
    df_ml_5 = pickle.load(myFile)

with open(f'../prem_clean_fixtures_and_dataframes/{df_10_saved_name}', 'rb') as myFile:
    df_ml_10 = pickle.load(myFile)

#缩放数据帧,使所有特征具有零均值和单位向量
df_ml_10 = scale_df(df_ml_10, list(range(14)), [14,15,16])
df_ml_5 = scale_df(df_ml_5, list(range(14)), [14,15,16])

x_10 = df_ml_10.drop(['Fixture ID', 'Team Result Indicator', 'Opponent Result Indicator'], axis=1)
y_10 = df_ml_10['Team Result Indicator']

x_5 = df_ml_5.drop(['Fixture ID', 'Team Result Indicator', 'Opponent Result Indicator'], axis=1)
y_5 = df_ml_5['Team Result Indicator']
```

（3）编写函数 svm_train() 实现 SVM 训练和支持向量机。代码如下：

```
def svm_train(df, print_result=True, print_result_label="):

    #创建特征矩阵
    x = df.drop(['Fixture ID', 'Team Result Indicator', 'Opponent Result Indicator'], axis=1)
    y = df['Team Result Indicator']

    #分为训练数据和测试数据
    x_train, x_test, y_train, y_test = train_test_split(x, y, test_size=0.2)

    #默认伽马值
    gamma = 1 / (14 * sum(x_train.var()))
    C = 1 / gamma

    #实例化SVM类
    clf = svm.SVC(kernel='rbf', C=3, probability=True)

    #训练模型
    clf.fit(x_train, y_train)

    if print_result:
        print(print_result_label)
        #训练数据
        train_data_score = round(clf.score(x_train, y_train) * 100, 1)
        print(f'Training data score = {train_data_score}%')
```

```
            #测试数据
            test_data_score = round(clf.score(x_test, y_test) * 100, 1)
            print(f'Test data score = {test_data_score}% \n')

        return clf, x_train, x_test, y_train, y_test

    ml_10_svm, x10_train, x10_test, y10_train, y10_test = svm_train(df_ml_10)
    ml_5_svm, x5_train, x5_test, y5_train, y5_test = svm_train(df_ml_5)
```

（4）编写函数 testing_c_parms() 用于测试参数 c。代码如下：

```
expo_iter = np.square(np.arange(0.1, 10, 0.1))

def testing_c_parms(df, iterable):
    training_score_li = []
    test_score_li = []
    for c in iterable:
        x = df.drop(['Fixture ID', 'Team Result Indicator', 'Opponent Result Indicator'], axis=1)
        y = df['Team Result Indicator']
        x_train, x_test, y_train, y_test = train_test_split(x, y, test_size=0.2, random_state = 1)
        clf = svm.SVC(kernel='rbf', C=c)
        clf.fit(x_train, y_train)
        train_data_score = round(clf.score(x_train, y_train) * 100, 1)
        test_data_score = round(clf.score(x_test, y_test) * 100, 1)
        training_score_li.append(train_data_score)
        test_score_li.append(test_data_score)
    return training_score_li, test_score_li

training_score_li, test_score_li = testing_c_parms(df_ml_10, expo_iter)

#c值约为3,可能比1更为理想
fig, ax = plt.subplots()
ax.plot(expo_iter, test_score_li)
```

（5）开始实现集成建模功能，分别使用相同的 SVM 算法和不同的输入数据来训练模型，这些特征大体上仍然相同，但是在不同的比赛数量中的平均值会不同，其中，df_ml_10 表示 10 场比赛，df_ml_5 表示 5 场比赛。代码如下：

```
#减少df_ml_5中的fixtures,使其仅包含df_ml_10中的features,并训练新数据集
df_ml_5_dropto10 = df_ml_5.drop(list(range(0,50)))
ml_5_to10_svm, x5_to10_train, x5_to10_test, y5_to10_train, y5_to10_test = svm_train
(df_ml_5_dropto10, print_result=False)

#独立使用两个df输入进行预测
y_pred_ml10 = ml_10_svm.predict(x10_test)
y_pred_ml5to10 = ml_5_to10_svm.predict(x10_test)

#独立地对每个数据集进行概率预测
```

```
pred_proba_ml10 = ml_10_svm.predict_proba(x10_test)
pred_proba_ml5_10 = ml_5_to10_svm.predict_proba(x10_test)

#组合独立概率和创建组合类预测
pred_proba_ml5and10 = (np.array(pred_proba_ml10) + np.array(pred_proba_ml5_10)) / 2.0
y_pred_ml5and10 = np.argmax(pred_proba_ml5and10, axis=1)

#准确度得分变量
y_pred_ml10_accuracy = round(accuracy_score(y10_test, y_pred_ml10), 3) * 100
y_pred_ml5to10_accuracy = round(accuracy_score(y10_test, y_pred_ml5to10), 3) * 100
y_pred_ml5and10_accuracy = round(accuracy_score(y10_test, y_pred_ml5and10), 3) * 100

print('ENSEMBLE MODEL TESTING')
print(f'Accuracy of df_10 alone = {y_pred_ml10_accuracy}%')
print(confusion_matrix(y10_test, y_pred_ml10), '\n')
print(f'Accuracy of df_5 alone = {y_pred_ml5to10_accuracy}%')
print(confusion_matrix(y10_test, y_pred_ml5to10), '\n')
print(f'Accuracy of df_5 and df_10 combined = {y_pred_ml5and10_accuracy}%')
print(confusion_matrix(y10_test, y_pred_ml5and10), '\n\n')
```

（6）输出显示不同比赛场次的模型评估和交叉验证功能。代码如下：

```
#交叉验证
skf = StratifiedKFold(n_splits=5, shuffle=True)

cv_score_av = round(np.mean(cross_val_score(ml_10_svm, x_10, y_10, cv=skf))*100,1)
print('Cross-Validation Accuracy Score ML10: ', cv_score_av, '%\n')

cv_score_av = round(np.mean(cross_val_score(ml_5_svm, x_5, y_5, cv=skf))*100,1)
print('Cross-Validation Accuracy Score ML5: ', cv_score_av, '%\n')
```

（7）绘制不同的可视化预测概率图。代码如下：

```
if pred_prob_plot_df10:
    fig = pred_proba_plot(ml_10_svm,
                          x_10,
                          y_10,
                          no_iter=50,
                          no_bins=36,
                          x_min=0.3,
                          classifier='Support Vector Machine (ml_10)')
    if save_pred_prob_plot_df10:
        fig.savefig('figures/ml_10_svm_pred_proba.png')

if pred_prob_plot_df5:
    fig = pred_proba_plot(ml_5_svm,
                          x_5,
                          y_5,
                          no_iter=50,
                          no_bins=36,
```

```
                            x_min=0.3,
                            classifier='Support Vector Machine (ml_5)')
    if save_pred_prob_plot_df5:
        fig.savefig('figures/ml_5_svm_pred_proba.png')
```

(8) 绘制不同的可视化混淆矩阵图。代码如下：

```
#打印混淆矩阵以获取交叉val结果
plot_cross_val_confusion_matrix(ml_10_svm,
                                x_10,
                                y_10,
                                display_labels=('team loses', 'draw', 'team wins'),
                                title='Support Vector Machine Confusion Matrix ML10',
                                cv=skf)
if save_conf_matrix_df10:
    plt.savefig('figures\ml_10_confusion_matrix_cross_val_svm.png')

plot_cross_val_confusion_matrix(ml_5_svm,
                                x_5,
                                y_5,
                                display_labels=('team loses', 'draw', 'team wins'),
                                title='Support Vector Machine Confusion Matrix ML5',
                                cv=skf)
if save_conf_matrix_df5:
    plt.savefig('figures\ml_5_confusion_matrix_cross_val_svm.png')
```

(9) 绘制不同的可视化学习曲线图。代码如下：

```
plot_learning_curve(ml_10_svm,
                    x_10,
                    y_10,
                    training_set_size=10,
                    x_max=240,
                    title='Learning Curve - Support Vector Machine DF_10',
                    leg_loc=1)
if save_learning_curve_df10:
    plt.savefig('figures\ml_10_svm_learning_curve.png')

plot_learning_curve(ml_5_svm,
                    x_5,
                    y_5,
                    training_set_size=10,
                    x_max=280,
                    title='Learning Curve - Support Vector Machine DF_5',
                    leg_loc=1)
if save_learning_curve_df5:
    plt.savefig('figures\ml_5_svm_learning_curve.png')
```

执行后会绘制不同的可视化评估图，例如 ML5 支持向量机混淆矩阵，如图 8-12 所示。

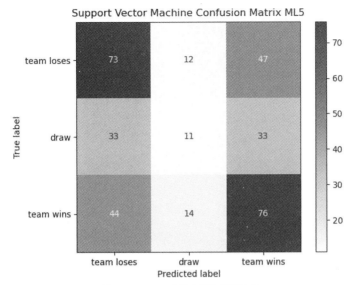

图 8-12　ML5 支持向量机混淆矩阵

DF_10 支持向量机学习曲线如图 8-13 所示。

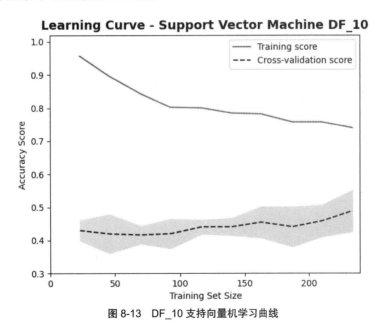

图 8-13　DF_10 支持向量机学习曲线

通过查看上述绘制的可视化矩阵，发现在预测平局时，所有三个模型都表现不佳，如图 8-14 所示。(a)随机森林模型的混淆矩阵为 50.8% 的准确度；(b)SVM 模型的混淆矩阵为 46.4% 的准确度；(c)最近邻模型的混淆矩阵为 51.5% 的准确度。

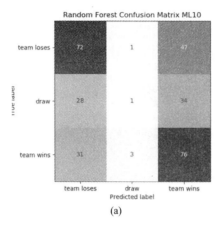

(a)

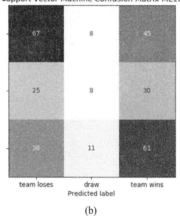

(b)

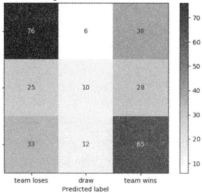

(c)

图 8-14 三个模型矩阵结果

8.7 Web 可视化

使用 Flask 创建一个 Web 页面, 在网页中展示新赛季英超联赛各队对阵时的预测结果的概率, 每场比赛的对阵结果有三种: 胜、平、负。

扫码观看本节视频讲解

8.7.1 获取预测数据

编写文件 python_anywhere_api_call_and_predictions.py, 功能是调用训练模型并对阵双方进行预测, 主要实现代码如下所示。

```
#----------------------------- 输入变量 -----------------------------

stats_dict_saved_name = '2019_2020_2021_prem_all_stats_dict.txt'
fixtures_saved_name = '2019_2020_2021_premier_league_fixtures_df.csv'
results_dict_saved_name = '2019_2020_2021_additional_stats_dict.txt'

#----------------------------- ADDITIONAL STATS -----------------------------

#加载我们已经生成的统计字典,并应用一些轻微的转换,以获得每个球队的df,该球队具有过去的结果和参与的团队
#这将用于我们网站上的"更多信息"下拉列表/可折叠列表

#----------加载数据----------

with open(f'../prem_clean_fixtures_and_dataframes/{stats_dict_saved_name}', 'rb') as myFile:
    game_stats = pickle.load(myFile)

fixtures_clean = pd.read_csv(f'../prem_clean_fixtures_and_dataframes/{fixtures_saved_name}')

#---------- 获取数据 ----------

teams_df = fixtures_clean.drop_duplicates(subset=['Home Team'])
teams_df = teams_df.drop(['Fixture ID', 'Game Date', 'Away Team', 'Away Team ID', 'Home Team Goals', 'Away Team Goals', 'Away Team Logo'], axis=1)
teams_df = teams_df.sort_values(by=['Home Team ID'])
teams_df = teams_df.reset_index(drop=True)
teams_df = teams_df.rename(columns={'Home Team ID': 'Team ID', 'Home Team': 'Team', 'Home Team Logo': 'Team Logo'})

def team_data(teams_df, team_id, return_data):
    '''
    return_data可以是以下三个变量之一: 'Venue', 'Team', 'Team Logo'
    '''
    team = teams_df.loc[teams_df['Team ID'] == team_id]
```

```python
        item = team[return_data]
        item = item.to_string(index=False)
        return item

test = team_data(teams_df, 50, 'Team')

#---------- DF 操纵----------

#为每支球队创建一个df,其中包含过去所有比赛的基本信息, 可以将其用作web应用程序中的显示
#然后将其放入字典中,密钥为球队的ID

#实例化字典和团队ID

results_dict = {}
teams = teams_df['Team ID']

for team in teams:
    df = pd.DataFrame(columns=['Fixture_ID', 'Date', 'Home_Team_ID','Away_Team_ID','Home_Team','Away_Team','Home_Team_Score','Away_Team_Score','Result','Home_Team_Logo','Away_Team_Logo'])

    dic = game_stats[team]
    fixture_id = list(dic.keys())

    if len(dic) == 0:
        nan_df = results_dict[33]
        nan_df['Home_Team'] = 'N/A'
        nan_df['Away_Team'] = 'N/A'
        nan_df['Home_Team_Score'] = 0
        nan_df['Away_Team_Score'] = 0
        nan_df['Fixture_ID'] = 'N/A'
        #nan_df['Date'] = 'N/A'
        nan_df['Home_Team_ID'] = 'N/A'
        nan_df['Away_Team_ID'] = 'N/A'
        nan_df['Home_Team_Logo'] = 'N/A'
        nan_df['Away_Team_Logo'] = 'N/A'
        nan_df['Result'] = 'N/A'
        results_dict[team] = nan_df
        continue
    game = dic[fixture_id[1]]

    date = []
    home_team_id = []
    away_team_id = []
    home_team = []
    away_team = []
    home_team_score = []
    away_team_score = []
    home_team_logo = []
```

```python
away_team_logo = []
results = []

for i, fix_id in enumerate(fixture_id):
    game = dic[fix_id]

    date.append(game['Game Date'].iloc[0])
    home_team_id.append(game['Team ID'].iloc[0])
    away_team_id.append(game['Team ID'].iloc[1])
    home_team_score.append(game['Goals'].iloc[0])
    away_team_score.append(game['Goals'].iloc[1])

df['Fixture_ID'] = fixture_id
df['Date'] = date
df['Home_Team_ID'] = home_team_id
df['Away_Team_ID'] = away_team_id
df['Home_Team_Score'] = home_team_score
df['Away_Team_Score'] = away_team_score

for i, home_team_ID in enumerate(df['Home_Team_ID']):
    home_team_str = team_data(teams_df, home_team_ID, 'Team')
    home_team_logo_str = team_data(teams_df, home_team_ID, 'Team Logo')
    home_team.append(home_team_str)
    home_team_logo.append(home_team_logo_str)

for i, away_team_ID in enumerate(df['Away_Team_ID']):
    away_team_str = team_data(teams_df, away_team_ID, 'Team')
    away_team_logo_str = team_data(teams_df, away_team_ID, 'Team Logo')
    away_team.append(away_team_str)
    away_team_logo.append(away_team_logo_str)

df['Home_Team'] = home_team
df['Away_Team'] = away_team
df['Home_Team_Logo'] = home_team_logo
df['Away_Team_Logo'] = away_team_logo

df = df.sort_values(by='Date', ascending=False)
df = df.reset_index(drop=True)

for i, home_team_ID in enumerate(df['Home_Team_ID']):

    if home_team_ID == team:
        home = True
    else:
        home = False

    home_score = df['Home_Team_Score'].iloc[i]
    away_score = df['Away_Team_Score'].iloc[i]
```

```python
            if home:
                if home_score > away_score:
                    result = 'W'
                elif home_score == away_score:
                    result = 'D'
                elif home_score < away_score:
                    result = 'L'

            if home==False:
                if home_score < away_score:
                    result = 'W'
                elif home_score == away_score:
                    result = 'D'
                elif home_score > away_score:
                    result = 'L'
            results.append(result)
        df['Result'] = results
        results_dict[team] = df


with open(f'../prem_clean_fixtures_and_dataframes/{results_dict_saved_name}', 'wb') as myFile:
    pickle.dump(results_dict, myFile)

print('\n', 'Script runtime:', round(((time.time()-start)/60), 2), 'minutes')
print(' ---------------- END ---------------- \n')
```

8.7.2　Flask Web 主页

每一个 Flask 项目都有一个启动主页，在本项目的 Flask Web 主页文件是 server.p，主要实现代码如下所示。

```python
from flask import Flask, render_template
import pickle
from datetime import datetime
app = Flask(__name__, static_url_path='/static')

#---------------------------------- FLASK ----------------------------------

with open('../predictions/pl_predictions.csv', 'rb') as myFile:
    pl_pred = pickle.load(myFile)

with open('../prem_clean_fixtures_and_dataframes/2019_2020_2021_additional_stats_dict.txt', 'rb') as myFile:
    additional_stats_dict = pickle.load(myFile)

#删除所有过去的预测(如果它们仍然存在于预测中)
```

```python
current_date = datetime.today().strftime('%Y-%m-%d')
for j in range(len(pl_pred)):
    game_date = pl_pred['Game Date'].loc[j]
    if game_date < current_date:
        pl_pred = pl_pred.drop([j], axis=0)
pl_pred = pl_pred.reset_index(drop=True)


#创建将在索引文件的for循环中使用的迭代器
max_display_games = 40
iterator_len = len(pl_pred) - 1
if iterator_len > max_display_games:
    iterator_len = max_display_games
iterator = range(iterator_len)

#创建将在索引文件的for循环中使用的迭代器,首先检查是否有足够的数据
max_additional_display_games = 5
dict_keys = list(additional_stats_dict.keys())
min_length = 100
for i in dict_keys:
    df_len = len(additional_stats_dict[i])
    if df_len < min_length:
        min_length = df_len
if max_additional_display_games > min_length:
    max_additional_display_games = min_length
iterator2 = range(max_additional_display_games)


@app.route('/')
def pass_game_1():
    return render_template('index.html',
                           pl_pred=pl_pred,
                           iterator=iterator,
                           iterator2=iterator2,
                           additional_stats_dict=additional_stats_dict)

if __name__ == '__main__':
    app.run(host = '0.0.0.0', port = 5000)
```

实时预测结果如图 8-15 所示。

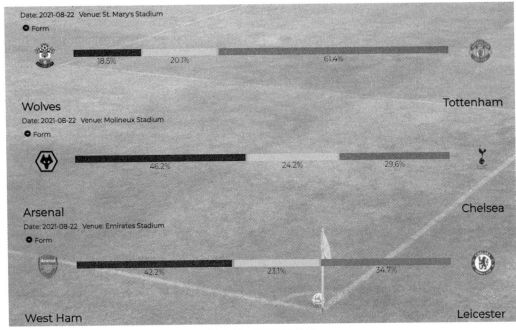

图 8-15 实时预测结果

第 9 章

AI考勤管理系统（face-recognition+Matplotlib+Django+Scikit-Learn+Dlib）

在新时代背景下，人工智能飞速发展，在商业办公领域，考勤打卡应用已经实现了无纸化处理。在本章的内容中，将详细介绍使用 Scikit-Learn 技术开发 AI 考勤管理系统的过程，讲解使用 face-recognition + Matplotlib + Django+Scikit-Learn+Dlib 实现大型人工智能项目的方法。

9.1 背景介绍

随着企业人事管理的日趋复杂和企业人员的增多,企业的考勤管理变得越来越复杂,有一个比较完善的考勤管理系统显得十分重要。考勤管理系统是使用计算机管理方式代替以前手工处理的工作。应用计算机技术和通信技术建立一个高效、无差错的考勤管理系统,能够有效地帮助企业实现"公正考勤,高效薪资",使企业的管理水平登上一个新的台阶。企业职工考勤管理系统,可用于各部门机构的职工考勤管理、查询、更新与维护,使用方便,易用性强,图形界面清晰明了。解决目前员工出勤管理问题,实现员工出勤信息和缺勤信息对企业领导透明,使管理人员及时掌握员工的情况,及时与员工沟通,提高管理效率。

扫码观看本节视频讲解

9.2 系统需求分析

需求分析是系统分析和软件设计阶段之间的桥梁,好的需求分析是项目成功的基石。一方面,需求分析以系统规格说明和项目规划作为分析活动的基本出发点,并从软件角度对它们进行检查与调整;另一方面,需求规格说明又是软件设计、实现、测试直至维护的主要基础。良好的分析活动有助于避免或尽早剔除早期错误,从而提高软件生产率,降低开发成本,改进软件质量。

扫码观看本节视频讲解

9.2.1 可行性分析

考勤管理是企业管理中非常重要的一环,作为公司主管考勤的人员能够通过考勤管理系统清楚地看到公司员工的签到时间、签离时间,以及是否迟到、早退等诸多信息,还能够通过所有员工的出勤记录比较来发现企业管理和员工作业方面的诸多问题。考勤管理系统更是员工工资及福利待遇方面重要的参考依据。

9.2.2 系统操作流程分析

(1)职工用户登录系统,上下班时进行签到考勤,经过系统验证通过后该员工签到成功。
(2)管理用户登录本系统,输入用户名和密码,系统进行验证,验证通过则进入程序主界面。在主界面对普通用户的信息进行录入,使用摄像头采集员工的人脸,然后通过机器学习技术创建学习模型。

9.2.3 系统模块设计

(1)登录验证模块。通过登录表单登录系统,整个系统分为管理员用户和普通员工用户。

（2）考勤打卡。普通用户登录系统后，可以分别实现在线上班打卡签到和下班打卡功能。

（3）添加新用户信息。管理员用户可以在后台添加新的员工信息，分别添加新员工的用户名和密码信息。

（4）采集照片。管理员用户可以在后台采集员工的照片，输入用户名，然后使用摄像头采集员工的图像。

（5）训练照片模型。使用机器学习技术训练采集到的员工照片，供员工打卡签到使用。

（6）考勤统计管理。使用可视化工具绘制员工的考勤数据，使用折线图统计最近两周每天到场的员工人数。本项目的功能模块如图 9-1 所示。

图 9-1　功能模块

9.3　系统配置

本系统是使用库 Django 实现的 Web 项目，在创建 Django Web 后会自动生成配置文件，开发者需要根据项目的需求设置这些配置文件。

9.3.1　Django 配置文件

扫码观看本节视频讲解

文件 settings.py 是 Django 项目的配置文件，主要用于设置整个 Django 项目所用到的程序文件和配置信息。在本项目中，需要设置本项目 SQLite3 数据库的名字为 db.sqlite3，并分别设置系统主页、登录页面和登录成功页面的 URL。文件 settings.py 的主要实现代码如下所示。

```
DATABASES = {
    'default': {
        'ENGINE': 'django.db.backends.sqlite3',
        'NAME': os.path.join(BASE_DIR, 'db.sqlite3'),
    }
}

STATIC_URL = '/static/'
CRISPY_TEMPLATE_PACK = 'bootstrap4'
LOGIN_URL='login'
LOGOUT_REDIRECT_URL = 'home'

LOGIN_REDIRECT_URL='dashboard'
```

9.3.2 路径导航文件

在 Django Web 项目中会自动创建路径导航文件 urls.py，在里面设置整个 Web 中所有页面对应的视图模块。本实例文件 urls.py 的主要实现代码如下所示。

```
urlpatterns = [
    path('admin/', admin.site.urls),
    path('', recog_views.home, name='home'),

    path('dashboard/', recog_views.dashboard, name='dashboard'),
    path('train/', recog_views.train, name='train'),
    path('add_photos/', recog_views.add_photos, name='add-photos'),

    path('login/',auth_views.LoginView.as_view(template_name='users/login.html'),
name='login'),
    path('logout/',auth_views.LogoutView.as_view(template_name='recognition/home.html'),
name='logout'),
    path('register/', users_views.register, name='register'),
    path('mark_your_attendance', recog_views.mark_your_attendance, name='mark-your-attendance'),
    path('mark_your_attendance_out', recog_views.mark_your_attendance_out,name=
'mark-your-attendance-out'),
    path('view_attendance_home', recog_views.view_attendance_home,name='view-
attendance-home'),
    path('view_attendance_date', recog_views.view_attendance_date,name='view-
attendance-date'),
    path('view_attendance_employee', recog_views.view_attendance_employee,
name='view-attendance-employee'),
    path('view_my_attendance', recog_views.view_my_attendance_employee_login,
name='view-my-attendance-employee-login'),
    path('not_authorised', recog_views.not_authorised, name='not-authorised')
]
```

9.4 用户注册和登录验证

为了提高开发效率，本项目使用库 Django 中的 django.contrib.auth 模块实现用户注册和登录验证功能。这样做的好处是减少代码编写量，节省开发时间。

扫码观看本节视频讲解

9.4.1 登录验证

根据文件 urls.py 中的如下代码可知，用户登录页面对应的模板文件是 login.html，此文件提供了用户登录表单，调用 django.contrib.auth 模块验证表单中的数据是否合法。

```
path('login/',auth_views.LoginView.as_view(template_name='users/login.html'),name='login'),
```

文件 login.html 的主要实现代码如下所示。

```
{% load static %}
{% load crispy_forms_tags %}

<!DOCTYPE html>
<html>
<head>

    <!-- Bootstrap CSS -->
      <link rel="stylesheet" href="https://maxcdn.bootstrapcdn.com/bootstrap/4.0.0/css/bootstrap.min.css" integrity="sha384-Gn5384xqQ1aoWXA+058RXPxPg6fy4IWvTNh0E263XmFcJlSAwiGgFAW/dAiS6JXm" crossorigin="anonymous">

    <style>
    body{
        background: url('{% static "recognition/img/bg_image.png"%}') no-repeat center center fixed;
        background-size: cover;

    }
    </style>

</head>
<body>

    <div class="col-lg-12" style="background: rgba(0,0,0,0.6);max-height: 20px; padding-top:1em;padding-bottom:3em;color:#fff;border-radius:10px;-webkit-box-shadow: 2px 2px 15px 0px rgba(0, 3, 0, 0.7);
    -moz-box-shadow:    2px 2px 15px 0px rgba(0, 3, 0, 0.7);
    box-shadow:         2px 2px 15px 0px rgba(0, 3, 0, 0.7); margin-left:auto; margin-right: auto; ">
```

```
    <a href="{% url 'home' %}"><h5 class="text-left"> Home</h5></a>
</div>

    <div class="col-lg-4" style="background: rgba(0,0,0,0.6);margin-top:300px; padding-top:1em;padding-bottom:3em;color:#fff;border-radius:10px;-webkit-box-shadow: 2px 2px 15px 0px rgba(0, 3, 0, 0.7);
    -moz-box-shadow:    2px 2px 15px 0px rgba(0, 3, 0, 0.7);
    box-shadow:         2px 2px 15px 0px rgba(0, 3, 0, 0.7); margin-left:auto; margin-right:auto; ">

    <form method="POST" >
        {% csrf_token %}
        <fieldset class="form-group">
            <legend class="border-bottom mb-4"> Log In </legend>
            {{form| crispy}}
        </fieldset>

        <div class="form-group">
            <button class="btn btn-outline-info" type="submit"> Login!</button>
        </div>
    </form>

</div>
```

用户登录验证表单页面的执行效果如图 9-2 所示。

图 9-2　用户登录验证表单页面

9.4.2　添加新用户

根据文件 urls.py 中的如下代码可知,新用户注册页面对应的功能模块是 users_views.register。

```
        path('register/', users_views.register, name='register'),
```

在文件 views.py 中，函数 register() 用于获取注册表单中的注册信息，实现新用户注册功能。文件 views.py 的主要实现代码如下所示。

```python
@login_required
def register(request):
    if request.user.username!='admin':
        return redirect('not-authorised')
    if request.method=='POST':
        form=UserCreationForm(request.POST)
        if form.is_valid():
            form.save() ###add user to database
            messages.success(request, f'Employee registered successfully!')
            return redirect('dashboard')

    else:
        form=UserCreationForm()
    return render(request,'users/register.html', {'form' : form})
```

在模板文件 register.html 中提供了注册表单功能，主要实现代码如下所示。

```html
<form method="POST" >
    {% csrf_token %}
    <fieldset class="form-group">
      <legend class="border-bottom mb-4"> Register New Employee </legend>
      {{form| crispy}}
    </fieldset>
    <div class="form-group">
      <button class="btn btn-outline-info" type="submit"> Register</button>
    </div>
  </form>
</div>
```

添加新用户表单页面的执行效果如图 9-3 所示。

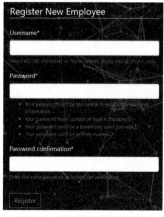

图 9-3　添加新用户表单页面

9.4.3 设计数据模型

在 Django Web 项目中，使用模型文件 models.py 设计项目中需要的数据库结构。因为本项目使用 django.contrib.auth 模块实现登录验证功能，所以在文件 models.py 中无须为会员用户设计数据库结构。模型文件 models.py 的主要实现代码如下所示。

```
from django.db import models
from django.contrib.auth.models import User

import datetime

class Present(models.Model):
    user=models.ForeignKey(User,on_delete=models.CASCADE)
    date = models.DateField(default=datetime.date.today)
    present=models.BooleanField(default=False)

class Time(models.Model):
    user=models.ForeignKey(User,on_delete=models.CASCADE)
    date = models.DateField(default=datetime.date.today)
    time=models.DateTimeField(null=True,blank=True)
    out=models.BooleanField(default=False)
```

通过上述代码设计了两个数据库表。
- Present：保存当前的打卡信息。
- Time：保存打卡的时间信息。

9.5 采集照片和机器学习

添加新的员工注册信息后，接下来需要采集员工的照片，然后使用 Scikit-Learn 将这些照片训练为机器学习模型，为员工的考勤打卡提供人脸识别和检测功能。

扫码观看本节视频讲解

9.5.1 设置采集对象

管理员用户成功登录系统后进入后台主页 http://127.0.0.1:8000/dashboard/，执行效果如图 9-4 所示。

管理员可以在后台采集员工的照片，单击 Add Photos 上面的"+"按钮后进入 http://127.0.0.1:8000/add_photos/ 页面，在此页面提供了如图 9-5 所示的表单，在表单中输入被采集对象的用户名。

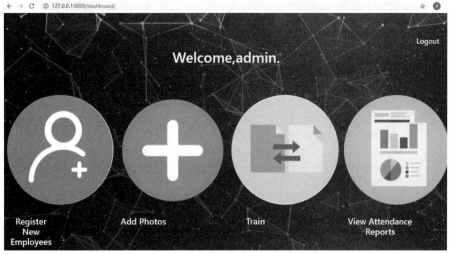

图 9-4　后台主页

图 9-5　输入被采集对象的用户名

根据文件 urls.py 中的如下代码可知，输入被采集对象用户名页面对应的视图模块是 recog_views.add_photos。

```
path('add_photos/', recog_views.add_photos, name='add-photos'),
```

在文件 views.py 中，视图函数 add_photos() 的功能是获取在表单中输入的用户名，验证输入的用户名是否在数据库中存在，如果存在则继续下一步的照片采集工作。函数 add_photos() 的具体实现代码如下所示。

```
@login_required
def add_photos(request):
    if request.user.username!='admin':
        return redirect('not-authorised')
    if request.method=='POST':
        form=usernameForm(request.POST)
```

```
                data = request.POST.copy()
                username=data.get('username')
                if username_present(username):
                    create_dataset(username)
                    messages.success(request, f'Dataset Created')
                    return redirect('add-photos')
                else:
                    messages.warning(request, f'No such username found. Please register employee first.')
                    return redirect('dashboard')
        else:

            form=usernameForm()
            return render(request,'recognition/add_photos.html', {'form' : form})
```

文件 add_photos.html 提供了输入被采集对象用户名的表单，主要实现代码如下所示。

```
<form method="POST" >
        {% csrf_token %}
        <fieldset class="form-group">
          <legend class="border-bottom mb-4"> Enter Username </legend>
          {{form| crispy}}
        </fieldset>

        <div class="form-group">
          <button class="btn btn-outline-info" type="submit"> Submit</button>
        </div>
      </form>
    </div>
<div class="col-lg-12" style="padding-top: 100px;">
 {% if messages %}
      {% for message in messages%}
      <div class="alert alert-{{message.tags}}" > {{message}}
      </div>
      {%endfor %}

    {%endif%}
  </div>
```

9.5.2 采集照片

在采集表单中输入用户名并单击 Submit 按钮后，打开当前电脑中的摄像头采集照片，然后采集照片中的人脸，并将人脸数据创建为 Dataset 文件。在文件 views.py 中，视图函数 create_dataset() 的功能是将采集的照片创建为 Dataset 文件，具体实现代码如下所示。

```
def create_dataset(username):
    id = username
```

```python
    if(os.path.exists('face_recognition_data/training_dataset/{}/'.format(id))==False):
        os.makedirs('face_recognition_data/training_dataset/{}/'.format(id))
    directory='face_recognition_data/training_dataset/{}/'.format(id)

    #检测人脸
    print("[INFO] Loading the facial detector")
    detector = dlib.get_frontal_face_detector()
    predictor = dlib.shape_predictor('face_recognition_data/shape_predictor_68_face_landmarks.dat')    #向形状预测器添加路径，稍后更改为相对路径
    fa = FaceAligner(predictor, desiredFaceWidth = 96)
    #从摄像头捕获图像并处理和检测人脸
    #初始化视频流
    print("[INFO] Initializing Video stream")
    vs = VideoStream(src=0).start()
    #time.sleep(2.0)

    #识别码,我们把id放在这里,并将id与一张脸一起存储,以便稍后我们可以识别它是谁的脸,将我们的数据集命名为计数器
    sampleNum = 0
    #一张一张地捕捉人脸,检测出人脸并显示在窗口上
    while(True):
        #拍摄图像,使用vs.read读取每一帧
        frame = vs.read()
        #调整每张图像的大小
        frame = imutils.resize(frame,width = 800)
        #返回的img是一张彩色图像,但是为了使分类器工作,我们需要一个灰度图像来转换
        gray_frame = cv2.cvtColor(frame, cv2.COLOR_BGR2GRAY)
        #存储人脸,检测当前帧中的所有图像,并返回图像中人脸的坐标和其他一些参数以获得准确的结果
        faces = detector(gray_frame,0)
        #在上面的"faces"变量中,可以有多个人脸,因此我们必须得到每个人脸,并在其周围绘制一个矩形
        for face in faces:
            print("inside for loop")
            (x,y,w,h) = face_utils.rect_to_bb(face)

            face_aligned = fa.align(frame,gray_frame,face)
            #每当程序捕捉到人脸时,我们都会把它写成一个文件夹
            #在捕获人脸之前,我们需要告诉脚本它是为谁的人脸创建的,我们需要一个标识符,这里我们称之为id
            #所以现抓到一张人脸后需要把它写进一个文件
            sampleNum = sampleNum+1
            #保存图像数据集,但只保存面部,裁剪掉其余的部分
            if face is None:
                print("face is none")
                continue

            cv2.imwrite(directory+'/'+str(sampleNum)+'.jpg', face_aligned)
            face_aligned = imutils.resize(face_aligned,width = 400)
            #cv2.imshow("Image Captured",face_aligned)
            # @params 矩形的初始点是 (x,y),终点是x的宽度和y的高度
```

```
        # #@params矩形的颜色
        # #@params矩形的厚度
        # @params
        cv2.rectangle(frame,(x,y),(x+w,y+h),(0,255,0),1)
        # 在继续下一个循环之前,设置50毫秒的暂停等待键
        cv2.waitKey(50)

    #在另一个窗口中显示图像,创建一个窗口,窗口名为"Face",图像为img
    cv2.imshow("Add Images",frame)
    #在关闭它之前,我们需要给出一个wait命令,否则opencv将无法工作,通过以下代码设置延迟
    1毫秒
    cv2.waitKey(1)
    #跳出循环
    if(sampleNum>300):
        break

#Stoping the videostream
vs.stop()
#销毁所有窗口
cv2.destroyAllWindows()
```

9.5.3 训练照片模型

在创建 Dataset 文件后单击后台主页中的 Train 图表按钮,使用机器学习技术 Scikit-Learn 训练 Dataset 数据集文件。根据文件 urls.py 中的如下代码可知,本项目通过 recog_views.train 模块训练 Dataset 数据集文件。

```
path('train/', recog_views.train, name='train'),
```

在视图文件 views.py 中,函数 predict() 的功能是实现预测处理,具体实现代码如下所示。

```
def predict(face_aligned,svc,threshold=0.7):
    face_encodings=np.zeros((1,128))
    try:
        x_face_locations=face_recognition.face_locations(face_aligned)
        faces_encodings=face_recognition.face_encodings(face_aligned,known_face_
locations=x_face_locations)
        if(len(faces_encodings)==0):
            return ([-1],[0])
    except:
        return ([-1],[0])

    prob=svc.predict_proba(faces_encodings)
    result=np.where(prob[0]==np.amax(prob[0]))
    if(prob[0][result[0]]<=threshold):
        return ([-1],prob[0][result[0]])
```

```python
    return (result[0],prob[0][result[0]])
```

在视图文件 views.py 中，函数 train() 的功能是训练数据集文件，具体实现代码如下所示。

```python
@login_required
def train(request):
    if request.user.username!='admin':
        return redirect('not-authorised')
    training_dir='face_recognition_data/training_dataset'

    count=0
    for person_name in os.listdir(training_dir):
        curr_directory=os.path.join(training_dir,person_name)
        if not os.path.isdir(curr_directory):
            continue
        for imagefile in image_files_in_folder(curr_directory):
            count+=1

    X=[]
    y=[]
    i=0

    for person_name in os.listdir(training_dir):
        print(str(person_name))
        curr_directory=os.path.join(training_dir,person_name)
        if not os.path.isdir(curr_directory):
            continue
        for imagefile in image_files_in_folder(curr_directory):
            print(str(imagefile))
            image=cv2.imread(imagefile)
            try:
                X.append((face_recognition.face_encodings(image)[0]).tolist())

                y.append(person_name)
                i+=1
            except:
                print("removed")
                os.remove(imagefile)

    targets=np.array(y)
    encoder=LabelEncoder()
    encoder.fit(y)
    y=encoder.transform(y)
    X1=np.array(X)
    print("shape: "+ str(X1.shape))
    np.save('face_recognition_data/classes.npy', encoder.classes_)
    svc=SVC(kernel='linear',probability=True)
    svc.fit(X1,y)
    svc_save_path="face_recognition_data/svc.sav"
    with open(svc_save_path, 'wb') as f:
```

```
        pickle.dump(svc,f)
    vizualize_Data(X1,targets)
    messages.success(request, f'Training Complete.')
    return render(request,"recognition/train.html")
```

训练完毕后会可视化展示训练结果，如图9-6所示，说明本项目在目前只是采集了两名员工的照片信息。

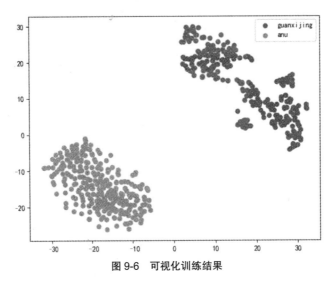

图 9-6　可视化训练结果

9.6　考勤打卡

员工登录系统主页后，可以分别实现在线上班打卡签到和下班打卡功能。在本节的内容中，将详细讲解实现考勤打卡功能的过程。

扫码观看本节视频讲解

9.6.1　上班打卡签到

在系统主页单击"Mark Your Attendance - In"上面的图标链接进入上班打卡页面 http://127.0.0.1:8000/mark_your_attendance，根据文件 urls.py 中的如下代码可知，考勤打卡页面功能是通过调用 recog_views.mark_your_attendance 模块实现的。

```
path('mark_your_attendance', recog_views.mark_your_attendance,name='mark-your-attendance'),
```

在视图文件 views.py 中，函数 mark_your_attendance() 的功能是采集摄像头中的人脸，根据前面训练的模型识别出是哪一名员工，然后实现考勤打卡功能，并将打卡信息添加到数据库中。函数 mark_your_attendance() 的具体实现代码如下所示。

```python
def mark_your_attendance(request):
    detector = dlib.get_frontal_face_detector()
    predictor = dlib.shape_predictor('face_recognition_data/shape_predictor_68_face_landmarks.dat')   #向形状预测器中添加路径,稍后更改为相对路径
    svc_save_path="face_recognition_data/svc.sav"
    with open(svc_save_path, 'rb') as f:
            svc = pickle.load(f)
    fa = FaceAligner(predictor, desiredFaceWidth = 96)
    encoder=LabelEncoder()
    encoder.classes_ = np.load('face_recognition_data/classes.npy')

    faces_encodings = np.zeros((1,128))
    no_of_faces = len(svc.predict_proba(faces_encodings)[0])
    count = dict()
    present = dict()
    log_time = dict()
    start = dict()
    for i in range(no_of_faces):
        count[encoder.inverse_transform([i])[0]] = 0
        present[encoder.inverse_transform([i])[0]] = False

    vs = VideoStream(src=0).start()
    sampleNum = 0
    while(True):
        frame = vs.read()
        frame = imutils.resize(frame ,width = 800)
        gray_frame = cv2.cvtColor(frame, cv2.COLOR_BGR2GRAY)
        faces = detector(gray_frame,0)

        for face in faces:
            print("INFO : inside for loop")
            (x,y,w,h) = face_utils.rect_to_bb(face)
            face_aligned = fa.align(frame,gray_frame,face)
            cv2.rectangle(frame,(x,y),(x+w,y+h),(0,255,0),1)
            (pred,prob)=predict(face_aligned,svc)
            if(pred!=[-1]):

                person_name=encoder.inverse_transform(np.ravel([pred]))[0]
                pred=person_name
                if count[pred] == 0:
                   start[pred] = time.time()
                   count[pred] = count.get(pred,0) + 1

                if count[pred] == 4 and (time.time()-start[pred]) > 1.2:
                    count[pred] = 0
                else:
                   present[pred] = True
                   log_time[pred] = datetime.datetime.now()
                   count[pred] = count.get(pred,0) + 1
```

```
                    print(pred, present[pred], count[pred])
                    cv2.putText(frame, str(person_name)+ str(prob), (x+6,y+h-6), cv2.FONT_
HERSHEY_SIMPLEX,0.5,(0,255,0),1)
                else:
                    person_name="unknown"
                    cv2.putText(frame, str(person_name), (x+6,y+h-6), cv2.FONT_HERSHEY_
SIMPLEX,0.5,(0,255,0),1)

                cv2.putText()
                #在继续下一个循环之前,设置一个50毫秒的暂停等待键
                cv2.waitKey(50)

        #在另一个窗口中显示图像
        #创建一个窗口,窗口名为"Face",图像为img
        cv2.imshow("Mark Attendance - In - Press q to exit",frame)
        #在关闭它之前,我们需要给出一个wait命令,否则opencv将无法工作,下面的参数表示
        延迟1毫秒
        cv2.waitKey(1)
        #停止循环
        key=cv2.waitKey(50) & 0xFF
        if(key==ord("q")):
            break
    #停止视频流
    vs.stop()

    #销毁所有窗体
    cv2.destroyAllWindows()
    update_attendance_in_db_in(present)
    return redirect('home')
```

9.6.2 下班打卡

在系统主页单击"Mark Your Attendance - Out"上面的图标链接进入下班打卡页面 http://127.0.0.1:8000/mark_your_attendance_out,根据文件 urls.py 中的如下代码可知,下班打卡页面功能是通过调用 recog_views.mark_your_attendance_out 模块实现的。

```
    path('mark_your_attendance_out', recog_views.mark_your_attendance_out,name='mark-
your-attendance-out'),
```

在视图文件 views.py 中,函数 mark_your_attendance_out() 的功能是采集摄像头中的人脸,根据前面训练的模型识别出是哪一名员工,然后实现下班打卡功能,并将打卡信息添加到数据库中。函数 mark_your_attendance_out() 的具体实现代码如下所示。

```
def mark_your_attendance_out(request):
    detector = dlib.get_frontal_face_detector()
    predictor = dlib.shape_predictor('face_recognition_data/shape_predictor_68_face_
```

```
landmarks.dat')    #向形状预测器添加路径，稍后更改为相对路径
        svc_save_path="face_recognition_data/svc.sav"

        with open(svc_save_path, 'rb') as f:
                svc = pickle.load(f)
        fa = FaceAligner(predictor, desiredFaceWidth = 96)
        encoder=LabelEncoder()
        encoder.classes_ = np.load('face_recognition_data/classes.npy')

        faces_encodings = np.zeros((1,128))
        no_of_faces = len(svc.predict_proba(faces_encodings)[0])
        count = dict()
        present = dict()
        log_time = dict()
        start = dict()
        for i in range(no_of_faces):
            count[encoder.inverse_transform([i])[0]] = 0
            present[encoder.inverse_transform([i])[0]] = False

        vs = VideoStream(src=0).start()
        sampleNum = 0
        while(True):
            frame = vs.read()
            frame = imutils.resize(frame ,width = 800)
            gray_frame = cv2.cvtColor(frame, cv2.COLOR_BGR2GRAY)
            faces = detector(gray_frame,0)

            for face in faces:
                print("INFO : inside for loop")
                (x,y,w,h) = face_utils.rect_to_bb(face)
                face_aligned = fa.align(frame,gray_frame,face)
                cv2.rectangle(frame,(x,y),(x+w,y+h),(0,255,0),1)

                (pred,prob)=predict(face_aligned,svc)
                if(pred!=[-1]):
                    person_name=encoder.inverse_transform(np.ravel([pred]))[0]
                    pred=person_name
                    if count[pred] == 0:
                      start[pred] = time.time()
                      count[pred] = count.get(pred,0) + 1
                    if count[pred] == 4 and (time.time()-start[pred]) > 1.5:
                        count[pred] = 0
                    else:
                      present[pred] = True
                      log_time[pred] = datetime.datetime.now()
                      count[pred] = count.get(pred,0) + 1
                      print(pred, present[pred], count[pred])
                    cv2.putText(frame, str(person_name)+ str(prob), (x+6,y+h-6), cv2.FONT_
HERSHEY_SIMPLEX,0.5,(0,255,0),1)
```

```
            else:
                person_name="unknown"
                cv2.putText(frame, str(person_name), (x+6,y+h-6), cv2.FONT_HERSHEY_
SIMPLEX,0.5,(0,255,0),1)

            #在另一个窗口中显示图像
            #将创建一个窗口,窗口名为"Face",图像为img
            cv2.imshow("Mark Attendance- Out - Press q to exit",frame)
            #在关闭它之前,我们需要给出一个wait命令,否则opencv将无法工作,下面的参数表示延迟1毫秒
            cv2.waitKey(1)
            key=cv2.waitKey(50) & 0xFF
            if(key==ord("q")):
                break
        vs.stop()

        cv2.destroyAllWindows()
        update_attendance_in_db_out(present)
        return redirect('home')
```

9.7 可视化考勤数据

管理登录系统后,可以在考勤统计管理页面查看员工的考勤信息。在本项目中,使用可视化工具绘制员工的考勤数据,使用折线图统计最近两周员工的考勤信息。

扫码观看本节视频讲解

9.7.1 统计最近两周员工的考勤数据

1. 视图函数

在后台主页单击"View Attendance Reports"上面的图表链接,在打开的网页 http://127.0.0.1:8000/view_attendance_home 中可以查看员工的考勤信息。根据文件 urls.py 中的如下代码可知,可视化考勤数据页面的功能是通过调用 recog_views.view_attendance_home 模块实现的。

```
path('view_attendance_home', recog_views.view_attendance_home,name='view-attendance-home'),
```

在视图文件 views.py 中,函数 view_attendance_home() 的功能是可视化展示员工的考勤数据,具体实现代码如下所示。

```
@login_required
def view_attendance_home(request):
    total_num_of_emp=total_number_employees()
    emp_present_today=employees_present_today()
    this_week_emp_count_vs_date()
    last_week_emp_count_vs_date()
```

```python
        return render(request,"recognition/view_attendance_home.html", {'total_num_of_emp' : 
total_num_of_emp, 'emp_present_today': emp_present_today})
```

在上述代码中用到了如下所示的 4 个函数。

(1) 函数 total_number_employees() 的功能是统计当前系统中员工的考勤信息，具体实现代码如下所示。

```python
def total_number_employees():
    qs=User.objects.all()
    return (len(qs) -1)
```

(2) 函数 employees_present_today() 的功能是统计今日打卡的员工数量，具体实现代码如下所示。

```python
def employees_present_today():
    today=datetime.date.today()
    qs=Present.objects.filter(date=today).filter(present=True)
    return len(qs)
```

(3) 函数 this_week_emp_count_vs_date() 的功能是统计本周每天员工的打卡信息，并绘制可视化折线图。具体实现代码如下所示。

```python
def this_week_emp_count_vs_date():
    today=datetime.date.today()
    some_day_last_week=today-datetime.timedelta(days=7)
    monday_of_last_week=some_day_last_week-datetime.timedelta(days=(some_day_last_week.isocalendar()[2] - 1))
    monday_of_this_week = monday_of_last_week + datetime.timedelta(days=7)
    qs=Present.objects.filter(date__gte=monday_of_this_week).filter(date__lte=today)
    str_dates=[]
    emp_count=[]
    str_dates_all=[]
    emp_cnt_all=[]
    cnt=0

    for obj in qs:
        date=obj.date
        str_dates.append(str(date))
        qs=Present.objects.filter(date=date).filter(present=True)
        emp_count.append(len(qs))
    while(cnt<5):
        date=str(monday_of_this_week+datetime.timedelta(days=cnt))
        cnt+=1
        str_dates_all.append(date)
        if(str_dates.count(date))>0:
            idx=str_dates.index(date)
            emp_cnt_all.append(emp_count[idx])
        else:
            emp_cnt_all.append(0)

    df=pd.DataFrame()
```

```python
df["date"]=str_dates_all
df["Number of employees"]=emp_cnt_all

sns.lineplot(data=df,x='date',y='Number of employees')
plt.savefig('./recognition/static/recognition/img/attendance_graphs/this_week/1.png')
plt.close()
```

（4）函数 last_week_emp_count_vs_date() 的功能是统计上一周每天员工的打卡信息，具体实现代码如下所示。

```python
def last_week_emp_count_vs_date():
    today=datetime.date.today()
    some_day_last_week=today-datetime.timedelta(days=7)
    monday_of_last_week=some_day_last_week-datetime.timedelta(days=(some_day_last_week.isocalendar()[2] - 1))
    monday_of_this_week = monday_of_last_week + datetime.timedelta(days=7)
    qs=Present.objects.filter(date__gte=monday_of_last_week).filter(date__lt=monday_of_this_week)
    str_dates=[]
    emp_count=[]
    str_dates_all=[]
    emp_cnt_all=[]
    cnt=0

    for obj in qs:
        date=obj.date
        str_dates.append(str(date))
        qs=Present.objects.filter(date=date).filter(present=True)
        emp_count.append(len(qs))
    while(cnt<5):
        date=str(monday_of_last_week+datetime.timedelta(days=cnt))
        cnt+=1
        str_dates_all.append(date)
        if(str_dates.count(date))>0:
            idx=str_dates.index(date)
            emp_cnt_all.append(emp_count[idx])
        else:
            emp_cnt_all.append(0)
    df=pd.DataFrame()
    df["date"]=str_dates_all
    df["emp_count"]=emp_cnt_all

    sns.lineplot(data=df,x='date',y='emp_count')
    plt.savefig('./recognition/static/recognition/img/attendance_graphs/last_week/1.png')
    plt.close()
```

2. 模板文件

编写模板文件 view_attendance_home.html，功能是调用上面的视图函数，使用曲线图可视化展示最近两周员工的考勤数据。主要实现代码如下所示。

```html
        <div class="collapse navbar-collapse" id="navbarNav">
          <ul class="navbar-nav">

            <li class="nav-item active">
              <a class="nav-link" href="{%url 'view-attendance-employee' %}">By Employee</a>
            </li>
              <li class="nav-item active">
                <a class="nav-link" href="{% url 'view-attendance-date' %}">By Date</a>
              </li>
                <li class="nav-item active" style="padding-left: 1440px">
                  <a class="nav-link" href="{% url 'dashboard' %}">Back to Admin Panel</a>
                </li>
          </ul>
        </div>
      </nav>


          <div class="card" style="margin-top: 2em; margin-left: 2em; margin-right: 2em; margin-bottom: 2em;">
            <div class="card-body">
      <h2> Today's Statistics </h2>
            <div class="row" style="margin-left: 12em">
      <div class="card" style="width: 20em; background-color: #338044; text-align : center; margin-left: 5em; margin-top: 5em; color: white;">
            <div class="card-body">
              <h5 class="card-title"> <b>Total Number Of Employees</b></h5>
              <p class="card-text" style="padding-top: 1em; font-size: 28px;"> <b>{{total_num_of_emp }}</b></p>
            </div>
          </div>
          <div class="card" style="width: 20em; background-color: #80335b; text-align : center; margin-left: 5em; margin-top: 5em; color: white;">
            <div class="card-body">
              <h5 class="card-title"> <b> Employees present today</b></h5>
              <p class="card-text" style="padding-top: 1em; font-size: 28px;"><b> {{emp_present_today }}</b></p>
            </div>
          </div>

        </div>
        </div>
        </div>


          <div class="card" style="margin-top: 2em; margin-left: 2em; margin-right: 2em; margin-bottom: 2em;">
            <div class="card-body">
          <div class="row">
      <div class="col-md-6">
      <h2> Last Week </h2>
```

```html
            <div class="card" style="width: 50em;">
             <img class="card-img-top" src="{% static 'recognition/img/attendance_graphs/last_week/1.png'%}" alt="Card image cap">
             <div class="card-body">
               <p class="card-text" style="text-align: center;">Number of employees present each day</p>
             </div>
            </div>

          </div>
          <div class="col-md-6">
            <h2> This Week </h2>
            <div class="card" style="width: 50em;">
             <img class="card-img-top" src="{% static 'recognition/img/attendance_graphs/this_week/1.png'%}" alt="Card image cap">
             <div class="card-body">
               <p class="card-text" style="text-align: center;">Number of employees present each day</p>
             </div>
            </div>
```

员工考勤数据可视化页面的执行效果如图9-7所示。

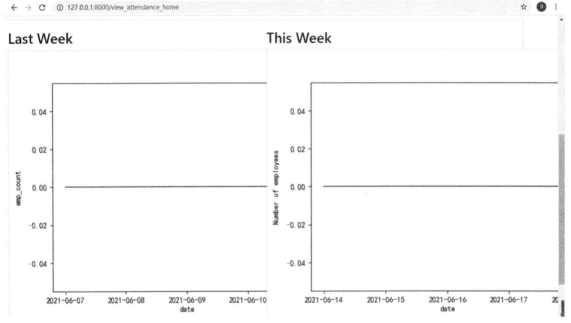

图 9-7　员工考勤数据可视化页面

9.7.2 查看本人在指定时间段内的考勤信息

1. 视图函数

普通员工登录系统后,单击"View My Attendance"上面的图标链接进入 http://127.0.0.1:8000/view_my_attendance 页面,如图9-8所示。

在此页面中可以查看本人在指定时间段内的考勤信息。根据文件 urls.py 中的如下代码可知,查看本人在指定时间段内的考勤数据的功能是通过调用 recog_views.view_my_attendance_employee_login 模块实现的。

```
path('view_my_attendance', recog_views.view_my_
attendance_employee_login,name='view-my-attendance-
employee-login'),
```

图9-8 选择时间段

在视图文件 views.py 中,函数 view_my_attendance_employee_login() 的功能是可视化展示本人在指定时间段内的考勤信息,具体实现代码如下所示。

```
@login_required
def view_my_attendance_employee_login(request):
    if request.user.username=='admin':
        return redirect('not-authorised')
    qs=None
    time_qs=None
    present_qs=None
    if request.method=='POST':
        form=DateForm_2(request.POST)
        if form.is_valid():
            u=request.user
            time_qs=Time.objects.filter(user=u)
            present_qs=Present.objects.filter(user=u)
            date_from=form.cleaned_data.get('date_from')
            date_to=form.cleaned_data.get('date_to')
            if date_to < date_from:
                messages.warning(request, f'Invalid date selection.')
                return redirect('view-my-attendance-employee-login')
            else:
                time_qs=time_qs.filter(date__gte=date_from).filter(date__lte=date_to).order_by('-date')
                present_qs=present_qs.filter(date__gte=date_from).filter(date__lte=date_to).order_by('-date')

                if (len(time_qs)>0 or len(present_qs)>0):
                    qs=hours_vs_date_given_employee(present_qs,time_qs,admin= False)
                    return render(request,'recognition/view_my_attendance_employee_login.html', {'form' : form, 'qs' :qs})
```

```
                else:
                        messages.warning(request, f'No records for selected duration.')
                        return redirect('view-my-attendance-employee-login')
        else:
                form=DateForm_2()
                return render(request,'recognition/view_my_attendance_employee_login.html',
{'form' : form, 'qs' :qs})
```

在上述代码中，调用函数 hours_vs_date_given_employee() 绘制在指定时间段内的考勤统计图，具体实现代码如下所示。

```
def hours_vs_date_given_employee(present_qs,time_qs,admin=True):
    register_matplotlib_converters()
    df_hours=[]
    df_break_hours=[]
    qs=present_qs
    for obj in qs:
        date=obj.date
        times_in=time_qs.filter(date=date).filter(out=False).order_by('time')
        times_out=time_qs.filter(date=date).filter(out=True).order_by('time')
        times_all=time_qs.filter(date=date).order_by('time')
        obj.time_in=None
        obj.time_out=None
        obj.hours=0
        obj.break_hours=0
        if (len(times_in)>0):
            obj.time_in=times_in.first().time
        if (len(times_out)>0):
            obj.time_out=times_out.last().time
        if(obj.time_in is not None and obj.time_out is not None):
            ti=obj.time_in
            to=obj.time_out
            hours=((to-ti).total_seconds())/3600
            obj.hours=hours
        else:
            obj.hours=0
        (check,break_hourss)= check_validity_times(times_all)
        if check:
            obj.break_hours=break_hourss

        else:
            obj.break_hours=0
        df_hours.append(obj.hours)
        df_break_hours.append(obj.break_hours)
        obj.hours=convert_hours_to_hours_mins(obj.hours)
        obj.break_hours=convert_hours_to_hours_mins(obj.break_hours)
    df = read_frame(qs)
    df["hours"]=df_hours
    df["break_hours"]=df_break_hours
```

```
print(df)
sns.barplot(data=df,x='date',y='hours')
plt.xticks(rotation='vertical')
rcParams.update({'figure.autolayout': True})
plt.tight_layout()
if(admin):
    plt.savefig('./recognition/static/recognition/img/attendance_graphs/hours_vs_date/1.png')
    plt.close()
else:
    plt.savefig('./recognition/static/recognition/img/attendance_graphs/employee_login/1.png')
    plt.close()
return qs
```

在上述代码中，如果当前登录用户是管理员，则统计在指定时间段内本人每天的上班时间。如果当前登录用户不是管理员，而是普通员工，则统计本人在指定时间段内的考勤信息，如图 9-9 所示。

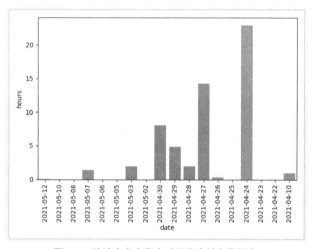

图 9-9　统计本人在指定时间段内的考勤信息

2. 模板文件

编写模板文件 view_attendance_date.html，功能是创建选择时间段的表单页面。主要实现代码如下所示。

```
<nav class="navbar navbar-expand-lg navbar-light bg-light">
<a class="navbar-brand" href="{%url 'view-attendance-home' %}">Attendance Dashboard</a>
    <button class="navbar-toggler" type="button" data-toggle="collapse" data-target="#navbarNav" aria-controls="navbarNav" aria-expanded="false" aria-label="Toggle navigation">
        <span class="navbar-toggler-icon"></span>
    </button>
```

```html
        <div class="collapse navbar-collapse" id="navbarNav">
          <ul class="navbar-nav">

            <li class="nav-item active">
              <a class="nav-link" href="{%url 'view-attendance-employee' %}">By Employee</a>
            </li>
                <li class="nav-item active">
                 <a class="nav-link" href="{% url 'view-attendance-date' %}">By Date</a>
            </li>
                 <li class="nav-item active" style="padding-left: 1440px">
                 <a class="nav-link" href="{% url 'dashboard' %}">Back to Admin Panel</a>
            </li>

          </ul>
        </div>
      </nav>

<div class="container">
  <div style="width: 400px">
 <form method="POST">
        {% csrf_token %}
        <fieldset class="form-group">
          <legend class="border-bottom mb-4"> Select Date </legend>
          {{form| crispy}}
        </fieldset>
        <div class="form-group">
          <button class="btn btn-outline-info" type="submit" value="Create"> Submit</button>
        </div>
      </form>
</div>
{% if qs %}
<table  class="table" style="margin-top: 5em;">
    <thead class="thead-dark">
    <tr>
        <th scope="col">Date</th>
        <th scope="col">Employee</th>
        <th scope="col">Present</th>
        <th scope="col">Time in</th>
        <th scope="col">Time out </th>
        <th scope="col">Hours </th>
        <th scope="col"> Break Hours </th>
    </tr>
</thead>
<tbody>
    {% for item in qs %}
    <tr>
        <td>{{ item.date }}</td>
        <td>{{ item.user.username}}</td>
     {% if item.present %}
```

```html
        <td> P </td>
        {% else %}
        <td> A </td>
        {% endif %}
        {% if item.time_in %}
        <td>{{ item.time_in }}</td>
        {% else %}
        <td> - </td>
        {% endif %}
          {% if item.time_out %}
        <td>{{ item.time_out }}</td>
        {% else %}
        <td> - </td>
        {% endif %}
            <td> {{item.hours}}</td>
            <td> {{item.break_hours}}</td>
    </tr>
    {% endfor %}
</tbody>
</table>
```

9.7.3 查看某员工在指定时间段内的考勤信息

1. 视图函数

管理员登录系统后，输入 URL 链接 http://127.0.0.1:8000/view_attendance_employee，如图 9-10 所示。在此页面中可以输入员工的名字和时间段，单击 Submit 按钮后可以查看某员工在指定时间段内的考勤信息。

在此页面中可以查看员工在指定时间段内的考勤信息。根据文件 urls.py 中的如下代码可知，查看某员工在指定时间段内的考勤信息的功能是通过调用 recog_views.view_attendance_employee 模块实现的。

```python
path('view_attendance_employee', recog_views.
view_attendance_employee,name='view-attendance-
employee'),
```

在视图文件 views.py 中，函数 view_attendance_employee() 的功能是可视化展示指定员工在指定时间段内的考勤信息，具体实现代码如下所示。

```python
@login_required
def view_attendance_employee(request):
    if request.user.username!='admin':
        return redirect('not-authorised')
```

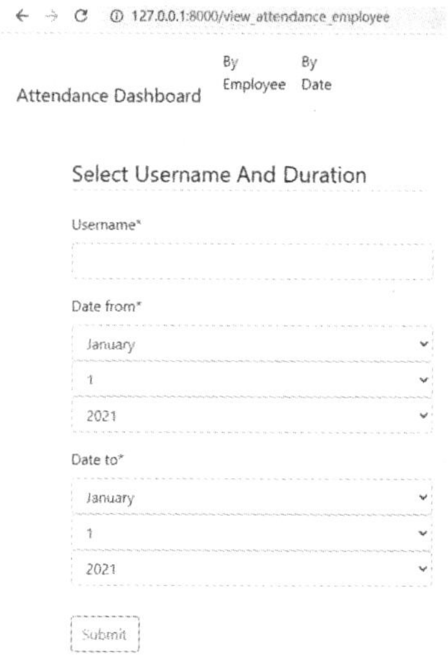

图 9-10 查看员工在某个时间段内的考勤信息

```python
            time_qs=None
            present_qs=None
            qs=None

            if request.method=='POST':
                form=UsernameAndDateForm(request.POST)
                if form.is_valid():
                    username=form.cleaned_data.get('username')
                    if username_present(username):
                        u=User.objects.get(username=username)
                        time_qs=Time.objects.filter(user=u)
                        present_qs=Present.objects.filter(user=u)
                        date_from=form.cleaned_data.get('date_from')
                        date_to=form.cleaned_data.get('date_to')
                        if date_to < date_from:
                           messages.warning(request, f'Invalid date selection.')
                           return redirect('view-attendance-employee')
                        else:
                           time_qs=time_qs.filter(date__gte=date_from).filter(date__lte=date_to).order_by('-date')
                           present_qs=present_qs.filter(date__gte=date_from).filter(date__lte=date_to).order_by('-date')

                           if (len(time_qs)>0 or len(present_qs)>0):
                               qs=hours_vs_date_given_employee(present_qs,time_qs,admin=True)
                               return render(request,'recognition/view_attendance_employee.html', {'form' : form, 'qs' :qs})
                           else:
                               #print("inside qs is None")
                               messages.warning(request, f'No records for selected duration.')
                               return redirect('view-attendance-employee')

                    else:
                        print("invalid username")
                        messages.warning(request, f'No such username found.')
                        return redirect('view-attendance-employee')
            else:
                    form=UsernameAndDateForm()
                    return render(request,'recognition/view_attendance_employee.html', {'form' : form, 'qs' :qs})
```

在上述代码中,需要使用视图函数 hours_vs_date_given_employee() 绘制柱状考勤统计图。

2. 模板文件

编写模板文件 view_attendance_employee.html,功能是创建设置员工用户名和选择时间段的表单页面。主要实现代码如下所示。

```html
<body>
  <nav class="navbar navbar-expand-lg navbar-light bg-light">
```

```html
        <a class="navbar-brand" href="{%url 'view-attendance-home' %}">Attendance Dashboard</a>
        <button class="navbar-toggler" type="button" data-toggle="collapse" data-target=
"#navbarNav" aria-controls="navbarNav" aria-expanded="false" aria-label="Toggle navigation">
            <span class="navbar-toggler-icon"></span>
        </button>
        <div class="collapse navbar-collapse" id="navbarNav">
          <ul class="navbar-nav">

            <li class="nav-item active">
              <a class="nav-link" href="{%url 'view-attendance-employee' %}">By Employee</a>
            </li>
              <li class="nav-item active">
              <a class="nav-link" href="{% url 'view-attendance-date' %}">By Date</a>
            </li>
                <li class="nav-item active" style="padding-left: 1440px">
              <a class="nav-link" href="{% url 'dashboard' %}">Back to Admin Panel</a>
            </li>

          </ul>
        </div>
      </nav>

<div class="container">
   <div style="width:400px;">

   <form method="POST">
        {% csrf_token %}
        <fieldset class="form-group">
            <legend class="border-bottom mb-4"> Select Username And Duration </legend>
            {{form| crispy}}
        </fieldset>

        <div class="form-group">
            <button class="btn btn-outline-info" type="submit"> Submit</button>
        </div>
    </form>

</div>

{%if qs%}
<table class="table"   style="margin-top: 5em;">
    <thead class="thead-dark">
    <tr>
        <th scope="col">Date</th>

        <th scope="col">Employee</th>
        <th scope="col">Present</th>
        <th scope="col">Time in</th>
        <th scope="col">Time out </th>
```

```html
                <th scope="col">Hours </th>
                <th scope="col"> Break Hours </th>
            </tr>
        </thead>
        <tbody>
            {% for item in qs %}
            <tr>
                <td>{{ item.date }}</td>
                <td>{{ item.user.username}}</td>

                {% if item.present %}
                <td> P </td>
                {% else %}
                <td> A </td>
                {% endif %}
                {% if item.time_in %}
                <td>{{ item.time_in }}</td>
                {% else %}
                <td> - </td>
                {% endif %}
                {% if item.time_out %}
                <td>{{ item.time_out }}</td>
                {% else %}
                <td> - </td>
                {% endif %}
                <td> {{item.hours}}</td>
                <td> {{item.break_hours}}</td>
            </tr>
            {% endfor %}
        </tbody>
    </table>

    <div class="card" style="margin-top: 5em; margin-bottom: 10em;">
      <img class="card-img-top" src="{% static 'recognition/img/attendance_graphs/hours_vs_date/1.png'%}" alt="Card image cap">
      <div class="card-body">
        <p class="card-text" style="text-align: center;">Number of hours worked each day.</p>
      </div>
    </div>
        {% endif %}
        {% if messages %}
        {% for message in messages%}
        <div class="alert alert-{{message.tags}}" > {{message}}
        </div>
        {%endfor %}
        {%endif%}
```

第 10 章

实时电影推荐系统（Scikit-Learn+Flask+Pandas）

推荐系统是指通过网站向用户提供商品、电影、新闻和音乐等信息的建议，帮助用户尽快找到自己感兴趣的信息。在本章的内容中，将使用 Scikit-Learn 开发一个实时电影推荐系统，详细介绍使用 Scikit-Learn 开发大型项目的知识。

10.1 系统介绍

推荐系统最早源于电子商务,在电子商务网站中向客户提供商品信息和建议,帮助用户决定应该购买什么产品,模拟销售人员帮助客户完成购买过程。个性化推荐能够根据用户的兴趣特点和购买行为推荐信息和商品。

扫码观看本节视频讲解

10.1.1 背景介绍

随着电子商务规模的不断扩大,商品个数和种类快速增长,顾客需要花费大量的时间才能找到自己想买的商品。这种浏览大量无关的信息和产品的过程无疑会使淹没在信息过载问题中的消费者不断流失。为了解决这些问题,个性化推荐系统应运而生。个性化推荐系统是建立在海量数据挖掘基础上的一种高级商务智能平台,以帮助电子商务网站为其顾客购物提供完全个性化的决策支持和信息服务。

互联网的出现和普及给用户带来了大量的信息,满足了用户在信息时代对信息的需求,但随着网络的迅速发展而带来的网上信息量的大幅增长,使得用户在面对大量信息时无法从中获得对自己真正有用的那部分信息,对信息的使用效率反而降低了,这就是所谓的信息超载(information overload)问题。

推荐系统是解决信息超载问题的一个非常有潜力的办法,它能够根据用户的信息需求、兴趣等,将用户感兴趣的信息、产品等推荐给用户。和搜索引擎相比,推荐系统通过研究用户的兴趣偏好,进行个性化计算,由系统发现用户的兴趣点,从而引导用户发现自己的信息需求。一个好的推荐系统不仅能为用户提供个性化的服务,还能和用户之间建立密切关系,让用户对推荐产生依赖。

推荐系统现已广泛应用于很多领域,其中,最典型并具有良好的发展和应用前景的领域就是电子商务领域。同时学术界对推荐系统的研究热度一直很高,逐步形成了一门独立的学科。

10.1.2 推荐系统和搜索引擎

当我们提到推荐引擎的时候,经常联想到的技术也便是搜索引擎。不必惊讶,因为这两者都是为了解决信息过载而提出的技术,一个问题,两个出发点。推荐系统和搜索引擎有共同的目标,即解决信息过载问题,但具体的做法因人而异。

搜索引擎更倾向于人们有明确的目的,可以将人们对于信息的寻求转换为精确的关键字,然后交给搜索引擎,最后返回给用户一系列列表,用户可以对这些返回结果进行反馈,但它会有马太效应的问题,即会造成越流行的东西随着搜索过程的迭代就越流行,使得那些越不流行的东西石沉大海。

而推荐引擎更倾向于人们没有明确的目的,或者说他们的目的是模糊的。通俗来讲,用户连自己都不知道他想要什么,这时候正是推荐引擎的用武之地。推荐系统通过用户的历史行为或者用户的兴趣偏好或者用户的人口统计学特征来送给推荐算法,然后推荐系统运用推荐算法来产生用户可能感兴趣的项目列表。其中,长尾理论(人们只关注曝光率高的项目,而忽略曝光率低的项目)可以很好地解释推荐系统的存在,试验表明,位于长尾位置的曝光率低的项目产生的利润不低于只销售曝光率高的项目的利润。

推荐系统正好可以给所有项目提供曝光的机会，以此来挖掘长尾项目的潜在利润。

如果说搜索引擎体现着马太效应的话，那么长尾理论则阐述了推荐系统所发挥的价值。

10.1.3 项目介绍

在本项目中，将提取训练过去几年在全球上映的电影信息，分别训练模型，提取用户情感数据。然后使用 Flask 开发一个 Web 网站，提供一个搜索表单供用户检索自己感兴趣的电影信息。当用户输入电影名字中的一个单词时，会自动弹出推荐的电影名字。即使用户输入的单词错误，也会提供推荐信息。选择某个推荐信息后，会在新页面中显示这部电影的相关信息，包括用户对这部电影的评价信息。

10.2 系统模块

本项目的系统模块结构如图 10-1 所示。

扫码观看本节视频讲解

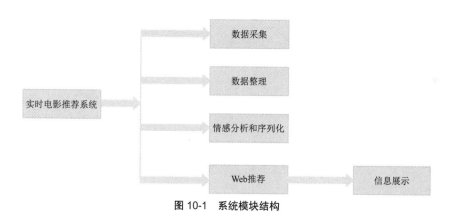

图 10-1 系统模块结构

10.3 数据采集和整理

本项目使用了多个数据集文件，包含 IMDB 5000 电影数据集、2018 年电影列表、2019 年电影列表和 2020 年电影列表。在本节的内容中，将介绍使用这些数据集提取整理数据并创建模型的知识。

扫码观看本节视频讲解

10.3.1 数据整理

编写文件 preprocessing 1.ipynb，基于数据集 movie_metadata.csv 整理里面的数据。文件 preprocessing 1.ipynb 的具体实现流程如下所示。

（1）导入头文件和数据集文件，查看前 10 条数据，代码如下：

```
import pandas as pd
import numpy as np
data = pd.read_csv('movie_metadata.csv')
data.head(10)
```

执行后会输出数据集中的前 10 条数据。

	color	director_name	num_critic_for_reviews	duration	director_facebook_likes	actor_3_facebook_likes	actor_2_name	actor_1_facebook_likes
0	Color	James Cameron	723.0	178.0	0.0	855.0	Joel David Moore	1000.0
1	Color	Gore Verbinski	302.0	169.0	563.0	1000.0	Orlando Bloom	40000.0
2	Color	Sam Mendes	602.0	148.0	0.0	161.0	Rory Kinnear	11000.0
3	Color	Christopher Nolan	813.0	164.0	22000.0	23000.0	Christian Bale	27000.0
4	NaN	Doug Walker	NaN	NaN	131.0	NaN	Rob Walker	131.0
5	Color	Andrew Stanton	462.0	132.0	475.0	530.0	Samantha Morton	640.0
6	Color	Sam Raimi	392.0	156.0	0.0	4000.0	James Franco	24000.0
7	Color	Nathan Greno	324.0	100.0	15.0	284.0	Donna Murphy	799.0
8	Color	Joss Whedon	635.0	141.0	0.0	19000.0	Robert Downey Jr.	26000.0
9	Color	David Yates	375.0	153.0	282.0	10000.0	Daniel Radcliffe	25000.0

（2）查看数据集矩阵的长度，代码如下：

```
data.shape
```

执行后会输出：

```
(5043, 28)
```

（3）返回数据集索引列表，代码如下：

```
data.columns
```

执行后会输出：

```
Index(['color', 'director_name', 'num_critic_for_reviews', 'duration',
       'director_facebook_likes', 'actor_3_facebook_likes', 'actor_2_name',
       'actor_1_facebook_likes', 'gross', 'genres', 'actor_1_name',
       'movie_title', 'num_voted_users', 'cast_total_facebook_likes',
       'actor_3_name', 'facenumber_in_poster', 'plot_keywords',
```

```
            'movie_imdb_link', 'num_user_for_reviews', 'language', 'country',
            'content_rating', 'budget', 'title_year', 'actor_2_facebook_likes',
            'imdb_score', 'aspect_ratio', 'movie_facebook_likes'],
           dtype='object')
```

(4）统计近年来的电影数量，代码如下：

```
import matplotlib.pyplot as plt
data.title_year.value_counts(dropna=False).sort_index().plot(kind='barh',figsize=(15,16))
plt.show()
```

执行效果如图 10-2 所示，由此可见，最早的电影数据是 1916 年。

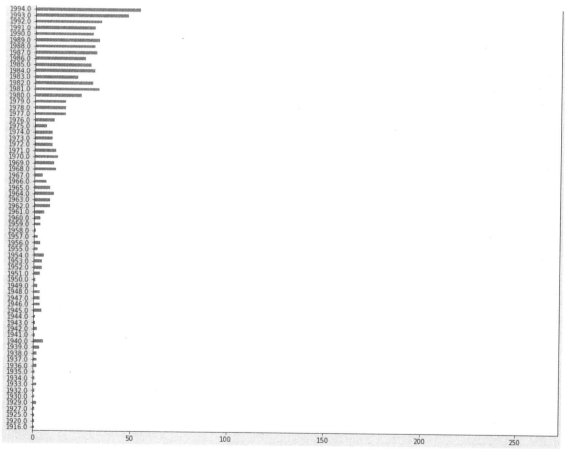

图 10-2　近年来的电影数量

（5）查看数据集中的前 10 条数据，只提取其中的几个字段，代码如下：

```
data = data.loc[:,['director_name',
'actor_1_name','actor_2_name','actor_3_name',
```

'genres','movie_title']]

执行后会输出:

	director_name	actor_1_name	actor_2_name	actor_3_name	genres	movie_title
0	James Cameron	CCH Pounder	Joel David Moore	Wes Studi	Action|Adventure|Fantasy|Sci-Fi	Avatar
1	Gore Verbinski	Johnny Depp	Orlando Bloom	Jack Davenport	Action|Adventure|Fantasy	Pirates of the Caribbean: At World's End
2	Sam Mendes	Christoph Waltz	Rory Kinnear	Stephanie Sigman	Action|Adventure|Thriller	Spectre
3	Christopher Nolan	Tom Hardy	Christian Bale	Joseph Gordon-Levitt	Action|Thriller	The Dark Knight Rises
4	Doug Walker	Doug Walker	Rob Walker	NaN	Documentary	Star Wars: Episode VII - The Force Awakens ...
5	Andrew Stanton	Daryl Sabara	Samantha Morton	Polly Walker	Action|Adventure|Sci-Fi	John Carter
6	Sam Raimi	J.K. Simmons	James Franco	Kirsten Dunst	Action|Adventure|Romance	Spider-Man 3
7	Nathan Greno	Brad Garrett	Donna Murphy	M.C. Gainey	Adventure|Animation|Comedy|Family|Fantasy|Musi...	Tangled
8	Joss Whedon	Chris Hemsworth	Robert Downey Jr.	Scarlett Johansson	Action|Adventure|Sci-Fi	Avengers: Age of Ultron
9	David Yates	Alan Rickman	Daniel Radcliffe	Rupert Grint	Adventure|Family|Fantasy|Mystery	Harry Potter and the Half-Blood Prince

（6）如果数据集中的某个值为空，则替换为"unknown"，代码如下：

```
data['actor_1_name'] = data['actor_1_name'].replace(np.nan, 'unknown')
data['actor_2_name'] = data['actor_2_name'].replace(np.nan, 'unknown')
data['actor_3_name'] = data['actor_3_name'].replace(np.nan, 'unknown')
data['director_name'] = data['director_name'].replace(np.nan, 'unknown')
data
```

执行后会输出:

	director_name	actor_1_name	actor_2_name	actor_3_name	genres	movie_title
0	James Cameron	CCH Pounder	Joel David Moore	Wes Studi	Action|Adventure|Fantasy|Sci-Fi	Avatar
1	Gore Verbinski	Johnny Depp	Orlando Bloom	Jack Davenport	Action|Adventure|Fantasy	Pirates of the Caribbean: At World's End
2	Sam Mendes	Christoph Waltz	Rory Kinnear	Stephanie Sigman	Action|Adventure|Thriller	Spectre
3	Christopher Nolan	Tom Hardy	Christian Bale	Joseph Gordon-Levitt	Action|Thriller	The Dark Knight Rises
4	Doug Walker	Doug Walker	Rob Walker	unknown	Documentary	Star Wars: Episode VII - The Force Awakens ...
...	...	...	...	...	...	...
5038	Scott Smith	Eric Mabius	Daphne Zuniga	Crystal Lowe	Comedy|Drama	Signed Sealed Delivered
5039	unknown	Natalie Zea	Valorie Curry	Sam Underwood	Crime|Drama|Mystery|Thriller	The Following
5040	Benjamin Roberds	Eva Boehnke	Maxwell Moody	David Chandler	Drama|Horror|Thriller	A Plague So Pleasant
5041	Daniel Hsia	Alan Ruck	Daniel Henney	Eliza Coupe	Comedy|Drama|Romance	Shanghai Calling
5042	Jon Gunn	John August	Brian Herzlinger	Jon Gunn	Documentary	My Date with Drew

5043 rows × 6 columns

（7）将"genres"列中的"|"替换为空格，代码如下：

```python
data['genres'] = data['genres'].str.replace('|', ' ')
data
```

执行后会输出：

	director_name	actor_1_name	actor_2_name	actor_3_name	genres	movie_title
0	James Cameron	CCH Pounder	Joel David Moore	Wes Studi	Action Adventure Fantasy Sci-Fi	Avatar
1	Gore Verbinski	Johnny Depp	Orlando Bloom	Jack Davenport	Action Adventure Fantasy	Pirates of the Caribbean: At World's End
2	Sam Mendes	Christoph Waltz	Rory Kinnear	Stephanie Sigman	Action Adventure Thriller	Spectre
3	Christopher Nolan	Tom Hardy	Christian Bale	Joseph Gordon-Levitt	Action Thriller	The Dark Knight Rises
4	Doug Walker	Doug Walker	Rob Walker	unknown	Documentary	Star Wars: Episode VII - The Force Awakens ...
...	...	...	...	...	...	...
5038	Scott Smith	Eric Mabius	Daphne Zuniga	Crystal Lowe	Comedy Drama	Signed Sealed Delivered
5039	unknown	Natalie Zea	Valorie Curry	Sam Underwood	Crime Drama Mystery Thriller	The Following
5040	Benjamin Roberds	Eva Boehnke	Maxwell Moody	David Chandler	Drama Horror Thriller	A Plague So Pleasant
5041	Daniel Hsia	Alan Ruck	Daniel Henney	Eliza Coupe	Comedy Drama Romance	Shanghai Calling
5042	Jon Gunn	John August	Brian Herzlinger	Jon Gunn	Documentary	My Date with Drew

5043 rows × 6 columns

（8）将 "movie_title" 列的数据变成小写，代码如下：

```python
data['movie_title'] = data['movie_title'].str.lower()
data['movie_title'][1]
```

执行后会输出：

"pirates of the caribbean: at world's end\xa0"

（9）删除 "movie_title" 结尾处的 null 终止字符，代码如下：

```python
data['movie_title'] = data['movie_title'].apply(lambda x : x[:-1])
data['movie_title'][1]
```

执行后会输出：

"pirates of the caribbean: at world's end"

（10）最后保存数据，代码如下：

```python
data.to_csv('data.csv',index=False)
```

10.3.2 电影详情数据

编写文件 preprocessing 2.ipynb，基于数据集文件 credits.csv 和 movies_metadata.csv 获取电影信息的详细数据。文件 preprocessing 2.ipynb 的具体实现流程如下所示。

（1）读取数据集文件 credits.csv 中的数据，代码如下：

```
credits = pd.read_csv('credits.csv')
credits
```

执行后会输出：

	cast	crew	id
0	[{'cast_id': 14, 'character': 'Woody (voice)', 'credit_id': '52fe4284c3...	[{'credit_id': '52fe4284c3a36847f8024f49', 'department': 'Directing', ...	862
1	[{'cast_id': 1, 'character': 'Alan Parrish', 'credit_id': '52fe44bfc3a3...	[{'credit_id': '52fe44bfc3a36847f80a7cd1', 'department': 'Production', ...	8844
2	[{'cast_id': 2, 'character': 'Max Goldman', 'credit_id': '52fe466a92514...	[{'credit_id': '52fe466a9251416c75077a89', 'department': 'Directing', ...	15602
3	[{'cast_id': 1, 'character': '"Savannah 'Vannah' Jackson"', 'credit_id':...	[{'credit_id': '52fe44779251416c91011acb', 'department': 'Directing', ...	31357
4	[{'cast_id': 1, 'character': 'George Banks', 'credit_id': '52fe44959251...	[{'credit_id': '52fe44959251416c75039ed7', 'department': 'Sound', 'gend...	11862
...	...	...	...
45471	[{'cast_id': 0, 'character': '', 'credit_id': '5894a909925141427e0079a5...	[{'credit_id': '5894a97d925141426c00818c', 'department': 'Directing', ...	439050
45472	[{'cast_id': 1002, 'character': 'Sister Angela', 'credit_id': '52fe4af1...	[{'credit_id': '52fe4af1c3a36847f81e9b15', 'department': 'Directing', ...	111109
45473	[{'cast_id': 6, 'character': 'Emily Shaw', 'credit_id': '52fe4776c3a368...	[{'credit_id': '52fe4776c3a368484e0c8387', 'department': 'Directing', ...	67758
45474	[{'cast_id': 2, 'character': '', 'credit_id': '52fe4ea59251416c7515d7d5...	[{'credit_id': '533bccebc3a36844cf0011a7', 'department': 'Directing', ...	227506
45475	[]	[{'credit_id': '593e676c92514105b702e68e', 'department': 'Directing', ...	461257

45476 rows × 3 columns

（2）读取数据集文件 movies_metadata.csv 中的内容，然后根据年份统计信息，代码如下：

```
meta = pd.read_csv('movies_metadata.csv')
meta['release_date'] = pd.to_datetime(meta['release_date'], errors='coerce')
meta['year'] = meta['release_date'].dt.year


meta['year'].value_counts().sort_index()
```

执行后会输出：

```
1874.0      1
1878.0      1
1883.0      1
1887.0      1
1888.0      2
           ...
2015.0   1905
2016.0   1604
2017.0    532
2018.0      5
2020.0      1
Name: year, Length: 135, dtype: int64
```

（3）因为在数据集中没有足够的 2018 年、2019 年和 2020 年的电影数据，因此只能获得 2017 年之前的电影信息。通过如下代码，预处理文件中 2017 年及其以前年份的电影数据。

```
new_meta = meta.loc[meta.year <= 2017,['genres','id','title','year']]
```

new_meta

执行后会输出：

	genres	id	title	year
0	[{'id': 16, 'name': 'Animation'}, {'id': 35, 'name': 'Comedy'}, {'id' ...	862	Toy Story	1995.0
1	[{'id': 12, 'name': 'Adventure'}, {'id': 14, 'name': 'Fantasy'}, {'id' ...	8844	Jumanji	1995.0
2	[{'id': 10749, 'name': 'Romance'}, {'id': 35, 'name': 'Comedy'}]	15602	Grumpier Old Men	1995.0
3	[{'id': 35, 'name': 'Comedy'}, {'id': 18, 'name': 'Drama'}, {'id': 1074...	31357	Waiting to Exhale	1995.0
4	[{'id': 35, 'name': 'Comedy'}]	11862	Father of the Bride Part II	1995.0
...	...	...	...	...
45460	[{'id': 18, 'name': 'Drama'}, {'id': 28, 'name': 'Action'}, {'id': 1074...	30840	Robin Hood	1991.0
45462	[{'id': 18, 'name': 'Drama'}]	111109	Century of Birthing	2011.0
45463	[{'id': 28, 'name': 'Action'}, {'id': 18, 'name': 'Drama'}, {'id': 53, ...	67758	Betrayal	2003.0
45464	[]	227506	Satan Triumphant	1917.0
45465	[]	461257	Queerama	2017.0

45370 rows × 4 columns

（4）在数据中添加两列"cast"和"crew"，代码如下：

```
new_meta['id'] = new_meta['id'].astype(int)
data = pd.merge(new_meta, credits, on='id')

pd.set_option('display.max_colwidth', 75)
data
```

执行后会输出：

	genres	id	title	year	cast	crew
0	[{'id': 16, 'name': 'Animation'}, {'id': 35, 'name': 'Comedy'}, {'id': ...	862	Toy Story	1995.0	[{'cast_id': 14, 'character': 'Woody (voice)', 'credit_id': '52fe4284c3...	[{'credit_id': '52fe4284c3a36847f8024f49', 'department': 'Directing', '...
1	[{'id': 12, 'name': 'Adventure'}, {'id': 14, 'name': 'Fantasy'}, {'id': ...	8844	Jumanji	1995.0	[{'cast_id': 1, 'character': 'Alan Parrish', 'credit_id': '52fe44bfc3a3...	[{'credit_id': '52fe44bfc3a36847f80a7cd1', 'department': 'Production', ...
2	[{'id': 10749, 'name': 'Romance'}, {'id': 35, 'name': 'Comedy'}]	15602	Grumpier Old Men	1995.0	[{'cast_id': 2, 'character': 'Max Goldman', 'credit_id': '52fe466a92514...	[{'credit_id': '52fe466a9251416c75077a89', 'department': 'Directing', '...
3	[{'id': 35, 'name': 'Comedy'}, {'id': 18, 'name': 'Drama'}, {'id': 1074...	31357	Waiting to Exhale	1995.0	[{'cast_id': 1, 'character': "Savannah 'Vannah' Jackson", 'credit_id': ...	[{'credit_id': '52fe44779251416c91011acb', 'department': 'Directing', '...
4	[{'id': 35, 'name': 'Comedy'}]	11862	Father of the Bride Part II	1995.0	[{'cast_id': 1, 'character': 'George Banks', 'credit_id': '52fe44959251...	[{'credit_id': '52fe44959251416c75039ed7', 'department': 'Sound', 'gend...
...	...	...	...	...	...	...
45440	[{'id': 18, 'name': 'Drama'}, {'id': 28, 'name': 'Action'}, {'id': 1074...	30840	Robin Hood	1991.0	[{'cast_id': 1, 'character': 'Sir Robert Hode', 'credit_id': '52fe44439...	[{'credit_id': '52fe44439251416c9100a899', 'department': 'Directing', '...
45441	[{'id': 18, 'name': 'Drama'}]	111109	Century of Birthing	2011.0	[{'cast_id': 1002, 'character': 'Sister Angela', 'credit_id': '52fe4af1...	[{'credit_id': '52fe4af1c3a36847f81e9b15', 'department': 'Directing', '...
45442	[{'id': 28, 'name': 'Action'}, {'id': 18, 'name': 'Drama'}, {'id': 53, ...	67758	Betrayal	2003.0	[{'cast_id': 6, 'character': 'Emily Shaw', 'credit_id': '52fe4776c3a368...	[{'credit_id': '52fe4776c3a368484e0c8387', 'department': 'Directing', '...
45443	[]	227506	Satan Triumphant	1917.0	[{'cast_id': 2, 'character': '', 'credit_id': '52fe4ea59251416c7515d7d5...	[{'credit_id': '533bccebc3a36844cf0011a7', 'department': 'Directing', '...
45444	[]	461257	Queerama	2017.0	[]	[{'credit_id': '593e676c92514105b702e68e', 'department': 'Directing', '...

（5）计算表达式节点或包含 Python 文本或容器显示的字符串，通过函数 make_genresList() 统计电影的类型，代码如下：

```python
import ast
data['genres'] = data['genres'].map(lambda x: ast.literal_eval(x))
data['cast'] = data['cast'].map(lambda x: ast.literal_eval(x))
data['crew'] = data['crew'].map(lambda x: ast.literal_eval(x))

def make_genresList(x):
    gen = []
    st = " "
    for i in x:
        if i.get('name') == 'Science Fiction':
            scifi = 'Sci-Fi'
            gen.append(scifi)
        else:
            gen.append(i.get('name'))
    if gen == []:
        return np.NaN
    else:
        return (st.join(gen))

data['genres_list'] = data['genres'].map(lambda x: make_genresList(x))

data['genres_list']
```

执行后会输出：

```
0               Animation Comedy Family
1               Adventure Fantasy Family
2                        Romance Comedy
3                 Comedy Drama Romance
4                                Comedy
                      ...
45440              Drama Action Romance
45441                             Drama
45442             Action Drama Thriller
45443                               NaN
45444                               NaN
Name: genres_list, Length: 45445, dtype: object
```

（6）编写自定义函数 get_actor1() 和 get_actor2() 获取 actor1 和 actor2 的信息，代码如下：

```python
def get_actor1(x):
    casts = []
    for i in x:
        casts.append(i.get('name'))
    if casts == []:
        return np.NaN
```

```
    else:
        return (casts[0])

data['actor_1_name'] = data['cast'].map(lambda x: get_actor1(x))

def get_actor2(x):
    casts = []
    for i in x:
        casts.append(i.get('name'))
    if casts == [] or len(casts)<=1:
        return np.NaN
    else:
        return (casts[1])data['actor_2_name'] = data['cast'].map(lambda x: get_actor2(x))

data['actor_2_name'] = data['cast'].map(lambda x: get_actor2(x))

data['actor_2_name']
```

执行后会输出：

```
0                  Tim Allen
1              Jonathan Hyde
2                Jack Lemmon
3             Angela Bassett
4              Diane Keaton
                 ...
45440            Uma Thurman
45441            Perry Dizon
45442           Adam Baldwin
45443       Nathalie Lissenko
45444                    NaN
Name: actor_2_name, Length: 45445, dtype: object
```

（7）编写自定义函数 get_actor3() 获取演员 3 的信息，代码如下：

```
def get_actor3(x):
    casts = []
    for i in x:
        casts.append(i.get('name'))
    if casts == [] or len(casts)<=2:
        return np.NaN
    else:
        return (casts[2])

data['actor_3_name'] = data['cast'].map(lambda x: get_actor3(x))

data['actor_3_name']
```

243

执行后会输出：

```
0              Don Rickles
1            Kirsten Dunst
2             Ann-Margret
3          Loretta Devine
4            Martin Short
                ...
45440        David Morrissey
45441         Hazel Orencio
45442         Julie du Page
45443          Pavel Pavlov
45444                   NaN
Name: actor_3_name, Length: 45445, dtype: object
```

（8）编写自定义函数 get_directors() 获取导演信息，代码如下：

```
def get_directors(x):
    dt = []
    st = " "
    for i in x:
        if i.get('job') == 'Director':
            dt.append(i.get('name'))
    if dt == []:
        return np.NaN
    else:
        return (st.join(dt))

data['director_name'] = data['crew'].map(lambda x: get_directors(x))

data['director_name']
```

执行后会输出：

```
0              John Lasseter
1               Joe Johnston
2              Howard Deutch
3             Forest Whitaker
4              Charles Shyer
                  ...
45440            John Irvin
45441             Lav Diaz
45442        Mark L. Lester
45443     Yakov Protazanov
45444         Daisy Asquith
Name: director_name, Length: 45445, dtype: object
```

（9）分别获取数据集中列 director_name、actor_1_name、actor_2_name、actor_3_name、genres_list 和

title 的信息，代码如下：

```
movie = data.loc[:,['director_name','actor_1_name','actor_2_name',
'actor_3_name','genres_list','title']]
movie
```

执行后会输出：

	director_name	actor_1_name	actor_2_name	actor_3_name	genres_list	title
0	John Lasseter	Tom Hanks	Tim Allen	Don Rickles	Animation Comedy Family	Toy Story
1	Joe Johnston	Robin Williams	Jonathan Hyde	Kirsten Dunst	Adventure Fantasy Family	Jumanji
2	Howard Deutch	Walter Matthau	Jack Lemmon	Ann-Margret	Romance Comedy	Grumpier Old Men
3	Forest Whitaker	Whitney Houston	Angela Bassett	Loretta Devine	Comedy Drama Romance	Waiting to Exhale
4	Charles Shyer	Steve Martin	Diane Keaton	Martin Short	Comedy	Father of the Bride Part II
...	...	...	...	...	...	...
45440	John Irvin	Patrick Bergin	Uma Thurman	David Morrissey	Drama Action Romance	Robin Hood
45441	Lav Diaz	Angel Aquino	Perry Dizon	Hazel Orencio	Drama	Century of Birthing
45442	Mark L. Lester	Erika Eleniak	Adam Baldwin	Julie du Page	Action Drama Thriller	Betrayal
45443	Yakov Protazanov	Iwan Mosschuchin	Nathalie Lissenko	Pavel Pavlov	NaN	Satan Triumphant
45444	Daisy Asquith	NaN	NaN	NaN	NaN	Queerama

45445 rows × 6 columns

（10）统计数据集中的数据数目，代码如下：

```
movie.isna().sum()
```

执行后会输出：

```
director_name    835
actor_1_name    2354
actor_2_name    3683
actor_3_name    4593
genres_list     2384
title              0
dtype: int64
```

（11）将"movie_title"列改为小写，然后打印输出定制的信息，代码如下：

```
movie = movie.rename(columns={'genres_list':'genres'})
movie = movie.rename(columns={'title':'movie_title'})

movie['movie_title'] = movie['movie_title'].str.lower()

movie['comb'] = movie['actor_1_name'] + ' ' + movie['actor_2_name'] + ' '+ movie
['actor_3_name'] + ' '+ movie['director_name'] +' ' + movie['genres']
```

```
movie
```

执行后会输出：

	director_name	actor_1_name	actor_2_name	actor_3_name	genres	movie_title	comb
0	John Lasseter	Tom Hanks	Tim Allen	Don Rickles	Animation Comedy Family	toy story	Tom Hanks Tim Allen Don Rickles John Lasseter Animation Comedy Family
1	Joe Johnston	Robin Williams	Jonathan Hyde	Kirsten Dunst	Adventure Fantasy Family	jumanji	Robin Williams Jonathan Hyde Kirsten Dunst Joe Johnston Adventure Fanta...
2	Howard Deutch	Walter Matthau	Jack Lemmon	Ann-Margret	Romance Comedy	grumpier old men	Walter Matthau Jack Lemmon Ann-Margret Howard Deutch Romance Comedy
3	Forest Whitaker	Whitney Houston	Angela Bassett	Loretta Devine	Comedy Drama Romance	waiting to exhale	Whitney Houston Angela Bassett Loretta Devine Forest Whitaker Comedy Dr...
4	Charles Shyer	Steve Martin	Diane Keaton	Martin Short	Comedy	father of the bride part ii	Steve Martin Diane Keaton Martin Short Charles Shyer Comedy
...	...	...	...	...	...	...	...
45438	Ben Rock	Monty Bane	Lucy Butler	David Grammer	Horror	the burkittsville 7	Monty Bane Lucy Butler David Grammer Ben Rock Horror
45439	Aaron Osborne	Lisa Boyle	Kena Land	Zaneta Polard	Sci-Fi	caged heat 3000	Lisa Boyle Kena Land Zaneta Polard Aaron Osborne Sci-Fi
45440	John Irvin	Patrick Bergin	Uma Thurman	David Morrissey	Drama Action Romance	robin hood	Patrick Bergin Uma Thurman David Morrissey John Irvin Drama Action Romance
45441	Lav Diaz	Angel Aquino	Perry Dizon	Hazel Orencio	Drama	century of birthing	Angel Aquino Perry Dizon Hazel Orencio Lav Diaz Drama
45442	Mark L. Lester	Erika Eleniak	Adam Baldwin	Julie du Page	Action Drama Thriller	betrayal	Erika Eleniak Adam Baldwin Julie du Page Mark L. Lester Action Drama Th...

39201 rows × 7 columns

（12）使用函数 drop_duplicates() 根据 "movie_title" 列实现去重处理，代码如下：

```
movie.drop_duplicates(subset ="movie_title", keep = 'last', inplace = True)
movie
```

执行后会输出：

	director_name	actor_1_name	actor_2_name	actor_3_name	genres	movie_title	comb
0	John Lasseter	Tom Hanks	Tim Allen	Don Rickles	Animation Comedy Family	toy story	Tom Hanks Tim Allen Don Rickles John Lasseter Animation Comedy Family
1	Joe Johnston	Robin Williams	Jonathan Hyde	Kirsten Dunst	Adventure Fantasy Family	jumanji	Robin Williams Jonathan Hyde Kirsten Dunst Joe Johnston Adventure Fanta...
2	Howard Deutch	Walter Matthau	Jack Lemmon	Ann-Margret	Romance Comedy	grumpier old men	Walter Matthau Jack Lemmon Ann-Margret Howard Deutch Romance Comedy
3	Forest Whitaker	Whitney Houston	Angela Bassett	Loretta Devine	Comedy Drama Romance	waiting to exhale	Whitney Houston Angela Bassett Loretta Devine Forest Whitaker Comedy Dr...
4	Charles Shyer	Steve Martin	Diane Keaton	Martin Short	Comedy	father of the bride part ii	Steve Martin Diane Keaton Martin Short Charles Shyer Comedy
...	...	...	...	...	...	...	...
45438	Ben Rock	Monty Bane	Lucy Butler	David Grammer	Horror	the burkittsville 7	Monty Bane Lucy Butler David Grammer Ben Rock Horror
45439	Aaron Osborne	Lisa Boyle	Kena Land	Zaneta Polard	Sci-Fi	caged heat 3000	Lisa Boyle Kena Land Zaneta Polard Aaron Osborne Sci-Fi
45440	John Irvin	Patrick Bergin	Uma Thurman	David Morrissey	Drama Action Romance	robin hood	Patrick Bergin Uma Thurman David Morrissey John Irvin Drama Action Romance
45441	Lav Diaz	Angel Aquino	Perry Dizon	Hazel Orencio	Drama	century of birthing	Angel Aquino Perry Dizon Hazel Orencio Lav Diaz Drama
45442	Mark L. Lester	Erika Eleniak	Adam Baldwin	Julie du Page	Action Drama Thriller	betrayal	Erika Eleniak Adam Baldwin Julie du Page Mark L. Lester Action Drama Th...

36341 rows × 7 columns

10.3.3 提取电影特征

编写文件 preprocessing 3.ipynb，功能是基于维基百科提取 2018 年电影数据的特征。文件 preprocessing 3.ipynb 的具体实现流程如下所示。

（1）设置要读取维基百科的数据信息的 URL，然后读取并显示数据。代码如下：

```
link = "https://en.wikipedia.org/wiki/List_of_American_films_of_2018"
df1 = pd.read_html(link, header=0)[2]
df2 = pd.read_html(link, header=0)[3]
df3 = pd.read_html(link, header=0)[4]
df4 = pd.read_html(link, header=0)[5]

df = df1.append(df2.append(df3.append(df4,ignore_index=True),ignore_index=True),ignore_index=True)

df
```

执行后会输出：

	Opening	Opening.1	Title	Production company	Cast and crew	Ref.
0	JANUARY	5	Insidious: The Last Key	Universal Pictures / Blumhouse Productions / S...	Adam Robitel (director); Leigh Whannell (scree...	[2]
1	JANUARY	5	The Strange Ones	Vertical Entertainment	Lauren Wolkstein (director); Christopher Radcl...	[3]
2	JANUARY	5	Stratton	Momentum Pictures	Simon West (director); Duncan Falconer, Warren...	[4]
3	JANUARY	10	Sweet Country	Samuel Goldwyn Films	Warwick Thornton (director); David Tranter, St...	[5]
4	JANUARY	12	The Commuter	Lionsgate / StudioCanal / The Picture Company	Jaume Collet-Serra (director); Byron Willinger...	[6]
...	...	...	...	...	...	...
263	DECEMBER	25	Holmes & Watson	Columbia Pictures / Gary Sanchez Productions	Etan Cohen (director/screenplay); Will Ferrell...	[162]
264	DECEMBER	25	Vice	Annapurna Pictures / Plan B Entertainment	Adam McKay (director/screenplay); Christian Ba...	[136]
265	DECEMBER	25	On the Basis of Sex	Focus Features	Mimi Leder (director); Daniel Stiepleman (scre...	[223]
266	DECEMBER	25	Destroyer	Annapurna Pictures	Karyn Kusama (director); Phil Hay, Matt Manfre...	[256]
267	DECEMBER	28	Black Mirror: Bandersnatch	Netflix	David Slade (director); Charlie Brooker (scree...	[257]

268 rows × 6 columns

（2）登录 themoviedb 官网，注册一个会员，然后申请一个 API 密钥，如图 10-3 所示。

图 10-3　themoviedb API 密钥

（3）编写自定义函数 get_genre() 获取 themoviedb 网站中的电影信息，在此需要用到 themoviedb API 密钥。代码如下：

```
from tmdbv3api import TMDb
import json
import requests
tmdb = TMDb()
tmdb.api_key = 'YOUR_API_KEY'

from tmdbv3api import Movie
tmdb_movie = Movie()
def get_genre(x):
    genres = []
    result = tmdb_movie.search(x)
    movie_id = result[0].id
    response = requests.get('https://api.themoviedb.org/3/movie/{}?api_key={}'.format(movie_id,tmdb.api_key))
    data_json = response.json()
    if data_json['genres']:
        genre_str = " "
        for i in range(0,len(data_json['genres'])):
            genres.append(data_json['genres'][i]['name'])
        return genre_str.join(genres)
    else:
        np.NaN

df['genres'] = df['Title'].map(lambda x: get_genre(str(x)))

df
```

执行后会输出：

	Opening	Opening.1	Title	Production company	Cast and crew	Ref.	genres
0	JANUARY	5	Insidious: The Last Key	Universal Pictures / Blumhouse Productions / S...	Adam Robitel (director); Leigh Whannell (scree...	[2]	Mystery Horror Thriller
1	JANUARY	5	The Strange Ones	Vertical Entertainment	Lauren Wolkstein (director); Christopher Radcl...	[3]	Thriller Drama
2	JANUARY	5	Stratton	Momentum Pictures	Simon West (director); Duncan Falconer, Warren...	[4]	Action Thriller
3	JANUARY	10	Sweet Country	Samuel Goldwyn Films	Warwick Thornton (director); David Tranter, St...	[5]	Drama History Western
4	JANUARY	12	The Commuter	Lionsgate / StudioCanal / The Picture Company	Jaume Collet-Serra (director); Byron Willinger...	[6]	Action Thriller
...	...	...	...	...	...	...	...
263	DECEMBER	25	Holmes & Watson	Columbia Pictures / Gary Sanchez Productions	Etan Cohen (director/screenplay); Will Ferrell...	[162]	Mystery Adventure Comedy Crime
264	DECEMBER	25	Vice	Annapurna Pictures / Plan B Entertainment	Adam McKay (director/screenplay); Christian Ba...	[136]	Thriller Science Fiction Action Adventure
265	DECEMBER	25	On the Basis of Sex	Focus Features	Mimi Leder (director); Daniel Stiepleman (scre...	[223]	Drama History
266	DECEMBER	25	Destroyer	Annapurna Pictures	Karyn Kusama (director); Phil Hay, Matt Manfre...	[256]	Thriller Crime Drama Action
267	DECEMBER	28	Black Mirror: Bandersnatch	Netflix	David Slade (director); Charlie Brooker (scree...	[257]	Science Fiction Mystery Drama Thriller TV Movie

268 rows × 7 columns

（4）只展示列 Title、Cast and crew 和 genres 中的内容，代码如下：

```
df_2018 = df[['Title','Cast and crew','genres']]
df_2018
```

执行后会输出：

	Title	Cast and crew	genres
0	Insidious: The Last Key	Adam Robitel (director); Leigh Whannell (scree...	Mystery Horror Thriller
1	The Strange Ones	Lauren Wolkstein (director); Christopher Radcl...	Thriller Drama
2	Stratton	Simon West (director); Duncan Falconer, Warren...	Action Thriller
3	Sweet Country	Warwick Thornton (director); David Tranter, St...	Drama History Western
4	The Commuter	Jaume Collet-Serra (director); Byron Willinger...	Action Thriller
...	...	...	...
263	Holmes & Watson	Etan Cohen (director/screenplay); Will Ferrell...	Mystery Adventure Comedy Crime
264	Vice	Adam McKay (director/screenplay); Christian Ba...	Thriller Science Fiction Action Adventure
265	On the Basis of Sex	Mimi Leder (director); Daniel Stiepleman (scre...	Drama History
266	Destroyer	Karyn Kusama (director); Phil Hay, Matt Manfre...	Thriller Crime Drama Action
267	Black Mirror: Bandersnatch	David Slade (director); Charlie Brooker (scree...	Science Fiction Mystery Drama Thriller TV Movie

268 rows × 3 columns

（5）分别编写自定义函数获取导演和演员的信息，代码如下：

```
def get_director(x):
    if " (director)" in x:
        return x.split(" (director)")[0]
    elif " (directors)" in x:
        return x.split(" (directors)")[0]
```

```
        else:
            return x.split(" (director/screenplay)")[0]

df_2018['director_name'] = df_2018['Cast and crew'].map(lambda x: get_director(x))

def get_actor1(x):
    return ((x.split("screenplay); ")[-1]).split(", ")[0])

df_2018['actor_1_name'] = df_2018['Cast and crew'].map(lambda x: get_actor1(x))

def get_actor2(x):
    if len((x.split("screenplay); ")[-1]).split(", ")) < 2:
        return np.NaN
    else:
        return ((x.split("screenplay); ")[-1]).split(", ")[1])
df_2018['actor_2_name'] = df_2018['Cast and crew'].map(lambda x: get_actor2(x))

def get_actor3(x):
    if len((x.split("screenplay); ")[-1]).split(", ")) < 3:
        return np.NaN
    else:
        return ((x.split("screenplay); ")[-1]).split(", ")[2])

df_2018['actor_3_name'] = df_2018['Cast and crew'].map(lambda x: get_actor3(x))
df_2018
```

执行后会输出：

	Title	Cast and crew	genres	director_name	actor_1_name	actor_2_name	actor_3_name
0	Insidious: The Last Key	Adam Robitel (director); Leigh Whannell (scree...	Mystery Horror Thriller	Adam Robitel	Lin Shaye	Angus Sampson	Leigh Whannell
1	The Strange Ones	Lauren Wolkstein (director); Christopher Radcl...	Thriller Drama	Lauren Wolkstein	Alex Pettyfer	James Freedson-Jackson	Emily Althaus
2	Stratton	Simon West (director); Duncan Falconer, Warren...	Action Thriller	Simon West	Dominic Cooper	Austin Stowell	Gemma Chan
3	Sweet Country	Warwick Thornton (director); David Tranter, St...	Drama History Western	Warwick Thornton	Bryan Brown	Sam Neill	NaN
4	The Commuter	Jaume Collet-Serra (director); Byron Willinger...	Action Thriller	Jaume Collet-Serra	Liam Neeson	Vera Farmiga	Patrick Wilson
...							
263	Holmes & Watson	Etan Cohen (director/screenplay); Will Ferrell...	Mystery Adventure Comedy Crime	Etan Cohen	Will Ferrell	John C. Reilly	Rebecca Hall
264	Vice	Adam McKay (director/screenplay); Christian Ba...	Thriller Science Fiction Action Adventure	Adam McKay	Christian Bale	Amy Adams	Steve Carell
265	On the Basis of Sex	Mimi Leder (director); Daniel Stiepleman (scre...	Drama History	Mimi Leder	Felicity Jones	Armie Hammer	Justin Theroux
266	Destroyer	Karyn Kusama (director); Phil Hay, Matt Manfre...	Thriller Crime Drama Action	Karyn Kusama	Nicole Kidman	Sebastian Stan	Toby Kebbell
267	Black Mirror: Bandersnatch	David Slade (director); Charlie Brooker (scree...	Science Fiction Mystery Drama Thriller TV Movie	David Slade	Fionn Whitehead	Will Poulter	Asim Chaudhry

268 rows × 7 columns

（6）将列"Title"重命名为"movie_title"，然后获取指定列的电影信息。代码如下：

```
df_2018 = df_2018.rename(columns={'Title':'movie_title'})
```

```
new_df18 = df_2018.loc[:,['director_name','actor_1_name','actor_2_name','actor_3_name',
'genres','movie_title']]
new_df18
```

执行后会输出:

	director_name	actor_1_name	actor_2_name	actor_3_name	genres	movie_title
0	Adam Robitel	Lin Shaye	Angus Sampson	Leigh Whannell	Mystery Horror Thriller	Insidious: The Last Key
1	Lauren Wolkstein	Alex Pettyfer	James Freedson-Jackson	Emily Althaus	Thriller Drama	The Strange Ones
2	Simon West	Dominic Cooper	Austin Stowell	Gemma Chan	Action Thriller	Stratton
3	Warwick Thornton	Bryan Brown	Sam Neill	NaN	Drama History Western	Sweet Country
4	Jaume Collet-Serra	Liam Neeson	Vera Farmiga	Patrick Wilson	Action Thriller	The Commuter
...	...	...	...	...	...	...
263	Etan Cohen	Will Ferrell	John C. Reilly	Rebecca Hall	Mystery Adventure Comedy Crime	Holmes & Watson
264	Adam McKay	Christian Bale	Amy Adams	Steve Carell	Thriller Science Fiction Action Adventure	Vice
265	Mimi Leder	Felicity Jones	Armie Hammer	Justin Theroux	Drama History	On the Basis of Sex
266	Karyn Kusama	Nicole Kidman	Sebastian Stan	Toby Kebbell	Thriller Crime Drama Action	Destroyer
267	David Slade	Fionn Whitehead	Will Poulter	Asim Chaudhry	Science Fiction Mystery Drama Thriller TV Movie	Black Mirror: Bandersnatch

268 rows × 6 columns

（7）修改其他演员的数值为"unknown"，将列"movie_title"的值转换为小写。代码如下：

```
new_df18['actor_2_name'] = new_df18['actor_2_name'].replace(np.nan, 'unknown')
new_df18['actor_3_name'] = new_df18['actor_3_name'].replace(np.nan, 'unknown')
new_df18['movie_title'] = new_df18['movie_title'].str.lower()
new_df18['comb'] = new_df18['actor_1_name'] + ' ' + new_df18['actor_2_name'] + ' '+ new_df18['actor_3_name'] + ' '+ new_df18['director_name'] +' ' + new_df18['genres']
```

执行后会输出:

	director_name	actor_1_name	actor_2_name	actor_3_name	genres	movie_title	comb
0	Adam Robitel	Lin Shaye	Angus Sampson	Leigh Whannell	Mystery Horror Thriller	insidious: the last key	Lin Shaye Angus Sampson Leigh Whannell Adam Ro...
1	Lauren Wolkstein	Alex Pettyfer	James Freedson-Jackson	Emily Althaus	Thriller Drama	the strange ones	Alex Pettyfer James Freedson-Jackson Emily Alt...
2	Simon West	Dominic Cooper	Austin Stowell	Gemma Chan	Action Thriller	stratton	Dominic Cooper Austin Stowell Gemma Chan Simon...
3	Warwick Thornton	Bryan Brown	Sam Neill	unknown	Drama History Western	sweet country	Bryan Brown Sam Neill unknown Warwick Thornton...
4	Jaume Collet-Serra	Liam Neeson	Vera Farmiga	Patrick Wilson	Action Thriller	the commuter	Liam Neeson Vera Farmiga Patrick Wilson Jaume ...
...	...	...	...	...	...	...	...
263	Etan Cohen	Will Ferrell	John C. Reilly	Rebecca Hall	Mystery Adventure Comedy Crime	holmes & watson	Will Ferrell John C. Reilly Rebecca Hall Etan ...
264	Adam McKay	Christian Bale	Amy Adams	Steve Carell	Thriller Science Fiction Action Adventure	vice	Christian Bale Amy Adams Steve Carell Adam McK...
265	Mimi Leder	Felicity Jones	Armie Hammer	Justin Theroux	Drama History	on the basis of sex	Felicity Jones Armie Hammer Justin Theroux Mim...
266	Karyn Kusama	Nicole Kidman	Sebastian Stan	Toby Kebbell	Thriller Crime Drama Action	destroyer	Nicole Kidman Sebastian Stan Toby Kebbell Kary...
267	David Slade	Fionn Whitehead	Will Poulter	Asim Chaudhry	Science Fiction Mystery Drama Thriller TV Movie	black mirror: bandersnatch	Fionn Whitehead Will Poulter Asim Chaudhry Dav...

268 rows × 7 columns

（8）按照上述流程从维基百科中提取 2019 年电影数据信息的特征，使用函数 isna() 实现去重处理，并保存处理后的数据集文件为 final_data.csv。代码如下：

```
final_df.isna().sum()

director_name    0
actor_1_name     0
actor_2_name     0
actor_3_name     0
genres           4
movie_title      0
comb             4
dtype: int64

final_df = final_df.dropna(how='any')

final_df.to_csv('final_data.csv',index=False)
```

10.4 情感分析和序列化操作

编写文件 sentiment.ipynb，功能是使用 pickle 模块实现数据序列化操作。通过 pickle 模块的序列化操作能够将程序中运行的对象信息保存到文件中去，永久存储；通过 pickle 模块的反序列化操作，能够从文件中创建上一次程序保存的对象。文件 sentiment.ipynb 的具体实现流程如下所示。

扫码观看本节视频讲解

（1）使用函数 nltk.download() 下载 stopwords，然后读取文件 reviews.txt 的内容，代码如下：

```
nltk.download("stopwords")
dataset = pd.read_csv('reviews.txt',sep = '\t', names =['Reviews','Comments'])
dataset
```

执行后会输出：

	Reviews	Comments
0	1	The Da Vinci Code book is just awesome.
1	1	this was the first clive cussler i've ever rea...
2	1	i liked the Da Vinci Code a lot.
3	1	i liked the Da Vinci Code a lot.
4	1	I liked the Da Vinci Code but it ultimately did...
...	...	...
6913	0	Brokeback Mountain was boring.
6914	0	So Brokeback Mountain was really depressing.
6915	0	As I sit here, watching the MTV Movie Awards, ...
6916	0	Ok brokeback mountain is such a horrible movie.
6917	0	Oh, and Brokeback Mountain was a terrible movie.

6918 rows × 2 columns

（2）使用函数 TfidfVectorizer() 将文本转换为可用作估算器输入的特征向量，然后将数据保存到文件 tranform.pkl 中，并计算准确度评分。代码如下：

```
topset = set(stopwords.words('english'))

vectorizer = TfidfVectorizer(use_idf = True,lowercase = True, strip_accents='ascii',
stop_words=stopset)

X = vectorizer.fit_transform(dataset.Comments)
y = dataset.Reviews
pickle.dump(vectorizer, open('tranform.pkl', 'wb'))

X_train, X_test, y_train, y_test = train_test_split(X, y, test_size=0.20, random_state=42)

clf = naive_bayes.MultinomialNB()
clf.fit(X_train,y_train)
accuracy_score(y_test,clf.predict(X_test))*100

clf = naive_bayes.MultinomialNB()
clf.fit(X,y)
```

执行后会分别输出准确度评分：

```
97.47109826589595
98.77167630057804
```

（3）最后将数据保存到文件 nlp_model.pkl，代码如下：

```
filename = 'nlp_model.pkl'
pickle.dump(clf, open(filename, 'wb'))
```

10.5　Web 端实时推荐

使用 Flask 编写前端程序，然后调用前面创建的文件 nlp_model.pkl 和 tranform.pkl 中的数据，在搜索电影时利用 Ajax 技术实现实时推荐功能，并通过 themoviedb API 展示要搜索电影的详细信息。

扫码观看本节视频讲解

10.5.1　Flask 启动页面

文件 main.py 是 Flask 的启动页面，功能是调用文件 nlp_model.pkl 和 tranform.pkl 中的数据，根据用户在表单中输入的数据提供实时推荐功能。文件 main.py 的主要实现代码如下所示。

```
#从磁盘加载nlp模型和tfidf矢量器
```

```python
filename = 'nlp_model.pkl'
clf = pickle.load(open(filename, 'rb'))
vectorizer = pickle.load(open('tranform.pkl','rb'))

#将字符串列表转换为列表 (eg. "["abc","def"]" to ["abc","def"])
def convert_to_list(my_list):
    my_list = my_list.split('","')
    my_list[0] = my_list[0].replace('["','')
    my_list[-1] = my_list[-1].replace('"]','')
    return my_list

#将数字列表转换为列表(eg. "[1,2,3]" to [1,2,3])
def convert_to_list_num(my_list):
    my_list = my_list.split(',')
    my_list[0] = my_list[0].replace("[","")
    my_list[-1] = my_list[-1].replace("]","")
    return my_list

def get_suggestions():
    data = pd.read_csv('main_data.csv')
    return list(data['movie_title'].str.capitalize())

app = Flask(__name__)

@app.route("/")
@app.route("/home")
def home():
    suggestions = get_suggestions()
    return render_template('home.html',suggestions=suggestions)


@app.route("/recommend",methods=["POST"])
def recommend():
    #从AJAX请求获取数据
    title = request.form['title']
    cast_ids = request.form['cast_ids']
    cast_names = request.form['cast_names']
    cast_chars = request.form['cast_chars']
    cast_bdays = request.form['cast_bdays']
    cast_bios = request.form['cast_bios']
    cast_places = request.form['cast_places']
    cast_profiles = request.form['cast_profiles']
    imdb_id = request.form['imdb_id']
    poster = request.form['poster']
    genres = request.form['genres']
    overview = request.form['overview']
    vote_average = request.form['rating']
    vote_count = request.form['vote_count']
    rel_date = request.form['rel_date']
```

```python
release_date = request.form['release_date']
runtime = request.form['runtime']
status = request.form['status']
rec_movies = request.form['rec_movies']
rec_posters = request.form['rec_posters']
rec_movies_org = request.form['rec_movies_org']
rec_year = request.form['rec_year']
rec_vote = request.form['rec_vote']

#获取自动完成的电影推荐
suggestions = get_suggestions()

#为每个需要转换为列表的字符串调用convert_to_list函数
rec_movies_org = convert_to_list(rec_movies_org)
rec_movies = convert_to_list(rec_movies)
rec_posters = convert_to_list(rec_posters)
cast_names = convert_to_list(cast_names)
cast_chars = convert_to_list(cast_chars)
cast_profiles = convert_to_list(cast_profiles)
cast_bdays = convert_to_list(cast_bdays)
cast_bios = convert_to_list(cast_bios)
cast_places = convert_to_list(cast_places)

#将字符串转换为列表
cast_ids = convert_to_list_num(cast_ids)
rec_vote = convert_to_list_num(rec_vote)
rec_year = convert_to_list_num(rec_year)

#将字符串呈现为python字符串
for i in range(len(cast_bios)):
    cast_bios[i] = cast_bios[i].replace(r'\n', '\n').replace(r'\"','\"')

for i in range(len(cast_chars)):
    cast_chars[i] = cast_chars[i].replace(r'\n', '\n').replace(r'\"','\"')

#将多个列表组合为一个字典,该字典可以传递到html文件,以便轻松处理该文件,并保留信息顺序
movie_cards = {rec_posters[i]: [rec_movies[i],rec_movies_org[i],rec_vote[i],rec_year[i]] for i in range(len(rec_posters))}

    casts = {cast_names[i]:[cast_ids[i], cast_chars[i], cast_profiles[i]] for i in range(len(cast_profiles))}

    cast_details = {cast_names[i]:[cast_ids[i], cast_profiles[i], cast_bdays[i], cast_places[i], cast_bios[i]] for i in range(len(cast_places))}

    #从IMDB站点获取用户评论的网页抓取
    sauce = urllib.request.urlopen('https://www.imdb.com/title/{}/reviews?ref_=tt_ov_rt'.format(imdb_id)).read()
    soup = bs.BeautifulSoup(sauce,'lxml')
```

```python
        soup_result = soup.find_all("div",{"class":"text show-more__control"})

        reviews_list = [] #审查清单
        reviews_status = [] #留言清单(good or bad)
        for reviews in soup_result:
            if reviews.string:
                reviews_list.append(reviews.string)
                #将评审传递给我们的模型
                movie_review_list = np.array([reviews.string])
                movie_vector = vectorizer.transform(movie_review_list)
                pred = clf.predict(movie_vector)
                reviews_status.append('Positive' if pred else 'Negative')

        #获取当前日期
        movie_rel_date = ""
        curr_date = ""
        if(rel_date):
            today = str(date.today())
            curr_date = datetime.strptime(today,'%Y-%m-%d')
            movie_rel_date = datetime.strptime(rel_date, '%Y-%m-%d')

        #将评论和审查合并到字典中
        movie_reviews = {reviews_list[i]: reviews_status[i] for i in range(len(reviews_list))}

        #将所有数据传递到html文件
        return render_template('recommend.html',title=title,poster=poster,overview=overview,
vote_average=vote_average,vote_count=vote_count,release_date=release_date,movie_rel_date=
movie_rel_date,curr_date=curr_date,runtime=runtime,status=status,genres=genres,movie_cards=movie_
cards,reviews=movie_reviews,casts=casts,cast_details=cast_details)

    if __name__ == '__main__':
        app.run(debug=True)
```

10.5.2 模板文件

在 Flask Web 项目中,使用模板文件实现前端功能。

(1) 本 Web 项目的主页是由模板文件 home.html 实现的,其功能是提供了一个表单供用户搜索电影,主要实现代码如下所示。

```html
<link rel= "stylesheet" type= "text/css" href= "{{ url_for('static',filename='style.css') }}">

<script type="text/javascript">
  var films = {{suggestions|tojson}};
  $(document).ready(function(){
    $("#myModal").modal('show');
  });
</script>
```

```html
    </head>

    <body id="content" style="font-family: 'Noto Sans JP', sans-serif;">
    <div class="body-content">
        <div class="ml-container" style="display: block;">
            <a href="https://github.com/kishan0725/The-Movie-Cinema" target="_blank" class="github-corner" title="View source on GitHub">
                <svg data-toggle="tooltip"
                    data-placement="left" width="80" height="80" viewBox="0 0 250 250"
                    style="fill:#e50914; color:#fff; position: fixed;z-index:100; top: 0; border: 0; right: 0;" aria-hidden="true">
                    <path d="M0,0 L115,115 L130,115 L142,142 L250,250 L250,0 Z"></path>
                    <path
                        d="M128.3,109.0 C113.8,99.7 119.0,89.6 119.0,89.6 C122.0,82.7 120.5,78.6 120.5,78.6 C119.2,72.0 123.4,76.3 123.4,76.3 C127.3,80.9 125.5,87.3 125.5,87.3 C122.9,97.1 130.6,101.9 134.4,103.2"
                        fill="currentColor" style="transform-origin: 130px 106px;" class="octo-arm"></path>
                    <path
                        d="M115.0,115.0 C114.9,115.1 118.7,116.5 119.8,115.4 L133.7,101.6 C136.9,99.2 139.9,98.4 142.2,98.6 C133.8,88.0 127.5,74.4 143.8,58.0 C148.5,53.4 154.0,51.2 159.7,51.0 C160.3,49.4 163.2,43.6 171.4,40.1 C171.4,40.1 176.1,42.5 178.8,56.2 C183.1,58.6 187.2,61.8 190.9,65.4 C194.5,69.0 197.7,73.2 200.1,77.6 C213.8,80.2 216.3,84.9 216.3,84.9 C212.7,93.1 206.9,96.0 205.4,96.6 C205.1,102.4 203.0,107.8 198.3,112.5 C181.9,128.9 168.3,122.5 157.7,114.1 C157.9,116.9 156.7,120.9 152.7,124.9 L141.0,136.5 C139.8,137.7 141.6,141.9 141.8,141.8 Z"
                        fill="currentColor" class="octo-body"></path>
                </svg>
            </a>
            <center><h1 class="app-title">电影推荐系统</h1></center>
            <div class="form-group shadow-textarea" style="margin-top: 30px;text-align: center;color: white;">
                <input type="text" name="movie" class="movie form-control" id="autoComplete"
                    autocomplete="off" placeholder="Enter the Movie Name" style="background-color: #ffffff;border-color:#ffffff;width: 60%;color: #181818" required="required" />
                <br>
            </div>

            <div class="form-group" style="text-align: center;">
                <button class="btn btn-primary btn-block movie-button" style="background-color: #e50914;text-align: center;border-color: #e50914;width:120px;" disabled="true" >Enter</button><br><br>
            </div>
        </div>

        <div id="loader" class="text-center">
        </div>
```

```html
        <div class="fail">
          <center><h3>很抱歉，您请求的电影不在我们的数据库中，请检查拼写或尝试其他电影!</h3></center>
        </div>

        <div class="results">
        <center>
          <h2 id="name" class="text-uppercase"></h2>
        </center>
        </div>

        <div class="modal fade" id="myModal" tabindex="-1" role="dialog" aria-labelledby=
"exampleModalLabel3" aria-hidden="true">
          <div class="modal-dialog modal-md" role="document">
            <div class="modal-content">
              <div class="modal-header" style="background-color: #e50914;color: white;">
                <h5 class="modal-title" id="exampleModalLabel3">Hey there!</h5>
                <button type="button" class="close" data-dismiss="modal" aria-label="Close">
                  <span aria-hidden="true" style="color: white">&times;</span>
                </button>
              </div>
              <div class="modal-body">
                <p>如果您正在寻找的电影在输入时没有获得实时推荐,请不要担心,只需输入电影名称并按
Enter键即可.即使你犯了一些打字错误,也可以得到推荐.</p>
              </div>
              <div class="modal-footer" style="text-align: center;">
                <button type="button" class="btn btn-secondary" data-dismiss="modal">知道了</button>
              </div>
            </div>
          </div>
        </div>

        <footer class="footer">
        <br/>
        <div class="social" style="margin-bottom: 8px">
          </div>
        </footer>
        </div>

        <script src="https://cdn.jsdelivr.net/npm/@tarekraafat/autocomplete.js@7.2.0/dist/
js/autoComplete.min.js"></script>
        <script type="text/javascript" src="{{url_for('static', filename='autocomplete.
js')}}"></script>

        <script type="text/javascript" src="{{url_for('static', filename='recommend.
js')}}"></script>
        <script src="https://cdnjs.cloudflare.com/ajax/libs/popper.js/1.12.9/umd/popper.min.
js" integrity="sha384-ApNbgh9B+Y1QKtv3Rn7W3mgPxhU9K/ScQsAP7hUibX39j7fakFPskvXusvfa0b4Q"
crossorigin="anonymous"></script>
        <script src="https://maxcdn.bootstrapcdn.com/bootstrap/4.0.0/js/bootstrap.min.
```

```
js" integrity="sha384-JZR6Spejh4U02d8jOt6vLEHfe/JQGiRRSQQxSfFWpi1MquVdAyjUar5+76PVCmYl" crossorigin="anonymous"></script>

    </body>
```

（2）编写模板文件 recommend.html，其功能是当用户在表单中输入某电影名称并按下 Enter 键后，会在此页面显示这部电影的详细信息。文件 recommend.html 的主要实现代码如下所示。

```
    <body id="content">
        <div class="results">
            <center>
              <h2 id="name" class="text-uppercase" style="font-family: 'Rowdies',cursive;">{{title}}</h2>
            </center>
        </div>
        <br/>

      <div id="mycontent">
        <div id="mcontent">
          <div class="poster-lg">
            <img class="poster" style="border-radius: 40px;margin-left: 90px;" height="400" width="250" src={{poster}}>
          </div>
          <div class="poster-sm text-center">
            <img class="poster" style="border-radius: 40px;margin-bottom: 5%;" height="400" width="250" src={{poster}}>
          </div>
          <div id="details">
            <br/>
            <h6 id="title" style="color:white;">电影名称:  {{title}}</h6>
            <h6 id="overview" style="color:white;max-width: 85%">简介: <br/><br/>      {{overview}}</h6>
            <h6 id="vote_average" style="color:white;">星级:  {{vote_average}}/10 ({{vote_count}} votes)</h6>
            <h6 id="genres" style="color:white;">类型:  {{genres}}</h6>
            <h6 id="date" style="color:white;">上映日期:  {{release_date}}</h6>
            <h6 id="runtime" style="color:white;">上映时长:  {{runtime}}</h6>
            <h6 id="status" style="color:white;">状态:  {{status}}</h6>
          </div>
        </div>
      </div>
      <br/>

    {% for name, details in cast_details.items() if not cast_details.hidden %}
      <div class="modal fade" id="{{details[0]}}" tabindex="-1" role="dialog" aria-labelledby="exampleModalLabel3" aria-hidden="true">
        <div class="modal-dialog modal-lg" role="document">
          <div class="modal-content">
```

```html
            <div class="modal-header" style="background-color: #e50914;color: white;">
              <h5 class="modal-title" id="exampleModalLabel3">{{name}}</h5>
              <button type="button" class="close" data-dismiss="modal" aria-label="Close">
                <span aria-hidden="true" style="color: white">&times;</span>
              </button>
            </div>

            <div class="modal-body">
              <img class="profile-pic" src="{{details[1]}}" alt="{{name}} - profile" style="width: 250px;height:400px;border-radius: 10px;" />
              <div style="margin-left: 20px">
                <p><strong>B生日:</strong> {{details[2]}} </p>
                <p><strong>出生地:</strong> {{details[3]}} </p>
                <p>
                  <p><strong>传记:</strong><p>
                  {{details[4]}}
                </p>
              </div>
            </div>
            <div class="modal-footer">
              <button type="button" class="btn btn-secondary" data-dismiss="modal">Close</button>
            </div>
          </div>
        </div>
      </div>
      {% endfor %}

      <div class="container">

        {% if casts|length > 1 %}
          <div class="movie" style="color: #E8E8E8;">
            <center>
              <h2 style="font-family: 'Rowdies', cursive;">演员列表</h2>
              <h5>(点击演员列表了解更多信息)</h5>
            </center>
          </div>



          <div class="movie-content">
            {% for name, details in casts.items() if not casts.hidden %}
              <div class="castcard card" style="width: 14rem;" title="Click to know more about {{name}}" data-toggle="modal" data-target="#{{details[0]}}">
                <div class="imghvr">
                  <img class="card-img-top cast-img" id="{{details[0]}}" height="360" width="240" alt="{{name}} - profile" src="{{details[2]}}">
                  <figcaption class="fig">
                    <button class="card-btn btn btn-danger"> Know More </button>
                  </figcaption>
                </div>
```

```html
                    <div class="card-body" style="font-family: 'Rowdies', cursive;font-size: 18px;">
                      <h5 class="card-title">{{name|upper}}</h5>
                      <h5 class="card-title" style="font-size: 18px"><span style="color:#756969;font-size: 18px;">AS {{details[1]|upper}}</span></h5>
                    </div>
                  </div>
                {% endfor %}
              </div>
            {% endif %}
            <br/>

            <center>
              {% if reviews %}
                <h2 style="font-family: 'Rowdies', cursive;color:white">USER REVIEWS</h2>
                <div class="col-md-12" style="margin: 0 auto; margin-top:25px;">
                  <table class="table table-bordered" bordercolor="white" style="color:white">
                    <thead>
                      <tr>
                        <th class="text-center" scope="col" style="width: 75%">评论</th>
                        <th class="text-center" scope="col">情感</th>
                      </tr>
                    </thead>

                    <tbody>
                      {% for review, status in reviews.items() if not reviews.hidden %}
                        <tr style="background-color:#e5091485;">
                          <td>{{review}}</td>
                          <td>
                            <center>
                              {{status}} :
                              {% if status =='Positive' %}
                                &#128515;
                              {% else %}
                                &#128534;
                              {% endif %}
                            </center>
                          </td>
                        </tr>
                      {% endfor %}
                    </tbody>
                  </table>
                </div>

              {% if (curr_date) and (movie_rel_date) %}
                {% elif curr_date < movie_rel_date %}
                  <div style="color:white;">
                    <h1 style="color:white"> This movie is not released yet. Stay tuned! </h1>
                  </div>
                {% else %}
```

```html
                <div style="color:white;">
                    <h1 style="color:white;"> Sorry, the reviews for this movie are not available! :( </h1>
                </div>
                {% endif %}
            {% else %}
                <div style="color:white;">
                    <h1 style="color:white;"> Sorry, the reviews for this movie are not available! :( </h1>
                </div>
            {% endif %}
        </center>
        <br/>


        {% if movie_cards|length > 1 %}

            <div class="movie" style="color: #E8E8E8;">
                <center><h2 style="font-family: 'Rowdies', cursive;">RECOMMENDED MOVIES FOR YOU</h2><h5>(Click any of the movies to get recommendation)</h5></center>
            </div>

            <div class="movie-content">
                {% for poster, details in movie_cards.items() if not movie_cards.hidden %}
                    <div class="card" style="width: 14rem;" title="{{details[1]}}" onclick="recommendcard(this)">
                        <div class="imghvr">
                            <img class="card-img-top" height="360" width="240" alt="{{details[0]}} - poster" src={{poster}}>
                            <div class="card-img-overlay" >
                                <span class="card-text" style="font-size:15px;background: #000000b8; color:white;padding:2px 5px;border-radius: 10px;"><span class="fa fa-star checked">  {{details[2]}}/10</span></span>
                            </div>
                            <div class=".card-img-overlay" style="position: relative;">
                                <span class="card-text" style="font-size:15px;position:absolute;bottom: 20px;left:15px;background: #000000b8;color:white;padding: 5px;border-radius: 10px;"> {{details[3]}}</span>
                            </div>
                            <figcaption class="fig">
                                <button class="card-btn btn btn-danger"> Click Me </button>
                            </figcaption>
                        </div>
                        <div class="card-body">
                            <h5 class="card-title" style="font-family: 'Rowdies', cursive;font-size: 17px;">{{details[0]|upper}}</h5>
                        </div>
                    </div>
                {% endfor %}
            </div>
        {% endif %}
```

```
<br/><br/><br/><br/>
  </div>
```

10.5.3 后端处理

在 Flask Web 项目中，除了使用主文件 main.py 实现后端处理功能外，还使用 JS 技术实现了后端功能。编写文件 recommend.js，其功能是调用 themoviedb API 实现实时推荐，并根据电影名获取这部电影的详细信息。文件 recommend.js 的具体实现流程如下所示。

（1）监听用户是否在电影搜索页面的文本框中输入内容，并监听是否单击 Enter 按钮。代码如下：

```javascript
$(function() {
    //按钮将被禁用,直到我们在输入字段中输入内容
    const source = document.getElementById('autoComplete');
    const inputHandler = function(e) {
      if(e.target.value==""){
        $('.movie-button').attr('disabled', true);
      }
      else{
        $('.movie-button').attr('disabled', false);
      }
    }
    source.addEventListener('input', inputHandler);

    $('.fa-arrow-up').click(function(){
      $('html, body').animate({scrollTop:0}, 'slow');
    });

    $('.app-title').click(function(){
      window.location.href = '/';
    })

    $('.movie-button').on('click',function(){
      var my_api_key = '你的API密钥';
      var title = $('.movie').val();
      if (title=="") {
        $('.results').css('display','none');
        $('.fail').css('display','block');
      }

      if (($('.fail').text() && ($('.footer').css('position') == 'absolute'))) {
        $('.footer').css('position', 'fixed');
      }

      else{
        load_details(my_api_key,title);
      }
```

```
  });
});
```

（2）编写函数 recommendcard()，以便在单击推荐的电影选项时调用此函数。代码如下：

```
function recommendcard(e){
  $("#loader").fadeIn();
  var my_api_key = '你的API密钥';
  var title = e.getAttribute('title');
  load_details(my_api_key,title);
}
```

（3）编写函数 load_details()，功能是从 API 获取电影的详细信息（基于电影名称）。代码如下：

```
function load_details(my_api_key,title){
  $.ajax({
    type: 'GET',
    url:'https://api.themoviedb.org/3/search/movie?api_key='+my_api_key+'&query='+title,
    async: false,
    success: function(movie){
      if(movie.results.length<1){
        $('.fail').css('display','block');
        $('.results').css('display','none');
        $("#loader").delay(500).fadeOut();
      }
      else if(movie.results.length==1) {
        $("#loader").fadeIn();
        $('.fail').css('display','none');
        $('.results').delay(1000).css('display','block');
        var movie_id = movie.results[0].id;
        var movie_title = movie.results[0].title;
        var movie_title_org = movie.results[0].original_title;
        get_movie_details(movie_id,my_api_key,movie_title,movie_title_org);
      }
      else{
        var close_match = {};
        var flag=0;
        var movie_id="";
        var movie_title="";
        var movie_title_org="";
        $("#loader").fadeIn();
        $('.fail').css('display','none');
        $('.results').delay(1000).css('display','block');
        for(var count in movie.results){
          if(title==movie.results[count].original_title){
            flag = 1;
            movie_id = movie.results[count].id;
            movie_title = movie.results[count].title;
            movie_title_org = movie.results[count].original_title;
            break;
```

```
            }
            else{
              close_match[movie.results[count].title] = similarity(title, movie.results
[count].title);
            }
          }
          if(flag==0){
              movie_title = Object.keys(close_match).reduce(function(a, b){ return close_
match[a] > close_match[b] ? a : b });
              var index = Object.keys(close_match).indexOf(movie_title)
              movie_id = movie.results[index].id;
              movie_title_org = movie.results[index].original_title;
          }
          get_movie_details(movie_id,my_api_key,movie_title,movie_title_org);
        }
    },
    error: function(error){
      alert('出错了 - '+error);
      $("#loader").delay(100).fadeOut();
    },
  });
}
```

（4）编写函数 similarity()，功能是使用距离参数 length 获取与请求的电影名称最接近的匹配。代码如下：

```
function similarity(s1, s2) {
  var longer = s1;
  var shorter = s2;
  if (s1.length < s2.length) {
    longer = s2;
    shorter = s1;
  }
  var longerLength = longer.length;
  if (longerLength == 0) {
    return 1.0;
  }
  return (longerLength - editDistance(longer, shorter)) / parseFloat(longerLength);
}

function editDistance(s1, s2) {
  s1 = s1.toLowerCase();
  s2 = s2.toLowerCase();

  var costs = new Array();
  for (var i = 0; i <= s1.length; i++) {
    var lastValue = i;
    for (var j = 0; j <= s2.length; j++) {
      if (i == 0)
```

```javascript
          costs[j] = j;
        else {
          if (j > 0) {
            var newValue = costs[j - 1];
            if (s1.charAt(i - 1) != s2.charAt(j - 1))
              newValue = Math.min(Math.min(newValue, lastValue),
                costs[j]) + 1;
            costs[j - 1] = lastValue;
            lastValue = newValue;
          }
        }
      }
      if (i > 0)
        costs[s2.length] = lastValue;
  }
  return costs[s2.length];
}
```

（5）编写函数 get_movie_details()，功能是根据电影 id 获取这部电影的所有详细信息。代码如下：

```javascript
function get_movie_details(movie_id,my_api_key,movie_title,movie_title_org) {
  $.ajax({
    type:'GET',
    url:'https://api.themoviedb.org/3/movie/'+movie_id+'?api_key='+my_api_key,
    success: function(movie_details){
      show_details(movie_details,movie_title,my_api_key,movie_id,movie_title_org);
    },
    error: function(error){
      alert("API Error! - "+error);
      $("#loader").delay(500).fadeOut();
    },
  });
}
```

（6）编写函数 show_details()，功能是将电影的详细信息传递给 Flask，以便使用 imdb id 显示和抓取这部电影的评论信息。代码如下：

```javascript
function show_details(movie_details,movie_title,my_api_key,movie_id,movie_title_org){
  var imdb_id = movie_details.imdb_id;
  var poster;
  if(movie_details.poster_path){
    poster = 'https://image.tmdb.org/t/p/original'+movie_details.poster_path;
  }
  else {
    poster = 'static/default.jpg';
  }
  var overview = movie_details.overview;
  var genres = movie_details.genres;
  var rating = movie_details.vote_average;
```

```
var vote_count = movie_details.vote_count;
var release_date = movie_details.release_date;
var runtime = parseInt(movie_details.runtime);
var status = movie_details.status;
var genre_list = [];
for (var genre in genres){
  genre_list.push(genres[genre].name);
}
var my_genre = genre_list.join(", ");
if(runtime%60==0){
  runtime = Math.floor(runtime/60)+" hour(s)"
}
else {
  runtime = Math.floor(runtime/60)+" hour(s) "+(runtime%60)+" min(s)"
}

//调用get_movie_cast以获取所查询电影的最佳演员阵容
movie_cast = get_movie_cast(movie_id,my_api_key);

//调用get_individual_cast以获取个人演员阵容的详细信息
ind_cast = get_individual_cast(movie_cast,my_api_key);

//调用get_Recommensions,从TMDB API获取给定电影id的推荐电影
recommendations = get_recommendations(movie_id, my_api_key);

details = {
    'title':movie_title,
    'cast_ids':JSON.stringify(movie_cast.cast_ids),
    'cast_names':JSON.stringify(movie_cast.cast_names),
    'cast_chars':JSON.stringify(movie_cast.cast_chars),
    'cast_profiles':JSON.stringify(movie_cast.cast_profiles),
    'cast_bdays':JSON.stringify(ind_cast.cast_bdays),
    'cast_bios':JSON.stringify(ind_cast.cast_bios),
    'cast_places':JSON.stringify(ind_cast.cast_places),
    'imdb_id':imdb_id,
    'poster':poster,
    'genres':my_genre,
    'overview':overview,
    'rating':rating,
    'vote_count':vote_count.toLocaleString(),
    'rel_date':release_date,
    'release_date':new Date(release_date).toDateString().split(' ').slice(1).join(' '),
    'runtime':runtime,
    'status':status,
    'rec_movies':JSON.stringify(recommendations.rec_movies),
    'rec_posters':JSON.stringify(recommendations.rec_posters),
    'rec_movies_org':JSON.stringify(recommendations.rec_movies_org),
    'rec_year':JSON.stringify(recommendations.rec_year),
```

```
      'rec_vote':JSON.stringify(recommendations.rec_vote)
    }

    $.ajax({
      type:'POST',
      data:details,
      url:"/recommend",
      dataType: 'html',
      complete: function(){
        $("#loader").delay(500).fadeOut();
      },
      success: function(response) {
        $('.results').html(response);
        $('#autoComplete').val('');
        $('.footer').css('position','absolute');
        if ($('.movie-content')) {
          $('.movie-content').after('<div class="gototop"><i title="Go to Top" class="fa fa-arrow-up"></i></div>');
        }
        $(window).scrollTop(0);
      }
    });
  }
```

（7）编写函数 get_individual_cast()，功能是获取某个演员的详细信息。代码如下：

```
function get_individual_cast(movie_cast,my_api_key) {
    cast_bdays = [];
    cast_bios = [];
    cast_places = [];
    for(var cast_id in movie_cast.cast_ids){
      $.ajax({
        type:'GET',
        url:'https://api.themoviedb.org/3/person/'+movie_cast.cast_ids[cast_id]+'?api_key='+my_api_key,
        async:false,
        success: function(cast_details){
          cast_bdays.push((new Date(cast_details.birthday)).toDateString().split(' ').slice(1).join(' '));
          if(cast_details.biography){
            cast_bios.push(cast_details.biography);
          }
          else {
            cast_bios.push("Not Available");
          }
          if(cast_details.place_of_birth){
            cast_places.push(cast_details.place_of_birth);
          }
          else {
```

```
          cast_places.push("Not Available");
        }
      }
    });
  }
  return {cast_bdays:cast_bdays,cast_bios:cast_bios,cast_places:cast_places};
}
```

（8）编写函数 get_movie_cast()，功能是获取所请求电影演员阵容的详细信息。代码如下：

```
function get_movie_cast(movie_id,my_api_key){
    cast_ids= [];
    cast_names = [];
    cast_chars = [];
    cast_profiles = [];
    top_10 = [0,1,2,3,4,5,6,7,8,9];
    $.ajax({
      type:'GET',
      url:"https://api.themoviedb.org/3/movie/"+movie_id+"/credits?api_key="+my_api_key,
      async:false,
      success: function(my_movie){
        if(my_movie.cast.length>0){
          if(my_movie.cast.length>=10){
            top_cast = [0,1,2,3,4,5,6,7,8,9];
          }
          else {
            top_cast = [0,1,2,3,4];
          }
          for(var my_cast in top_cast){
            cast_ids.push(my_movie.cast[my_cast].id);
            cast_names.push(my_movie.cast[my_cast].name);
            cast_chars.push(my_movie.cast[my_cast].character);
            if(my_movie.cast[my_cast].profile_path){
              cast_profiles.push("https://image.tmdb.org/t/p/original"+my_movie.cast[my_cast].profile_path);
            }
            else {
              cast_profiles.push("static/default.jpg");
            }
          }
        }
      },
      error: function(error){
        alert("出错了！ - "+error);
        $("#loader").delay(500).fadeOut();
      }
    });
    return {cast_ids:cast_ids,cast_names:cast_names,cast_chars:cast_chars,cast_profiles:
```

```
cast_profiles};
    }
```

（9）编写函数 get_recommendations()，功能是获得实时推荐的电影信息。代码如下：

```javascript
function get_recommendations(movie_id, my_api_key) {
    rec_movies = [];
    rec_posters = [];
    rec_movies_org = [];
    rec_year = [];
    rec_vote = [];

    $.ajax({
        type: 'GET',
        url: "https://api.themoviedb.org/3/movie/"+movie_id+"/recommendations?api_key="+my_api_key,
        async: false,
        success: function(recommend) {
            for(var recs in recommend.results) {
                rec_movies.push(recommend.results[recs].title);
                rec_movies_org.push(recommend.results[recs].original_title);
                rec_year.push(new Date(recommend.results[recs].release_date).getFullYear());
                rec_vote.push(recommend.results[recs].vote_average);
                if(recommend.results[recs].poster_path){
                    rec_posters.push("https://image.tmdb.org/t/p/original"+recommend.results[recs].poster_path);
                }
                else {
                    rec_posters.push("static/default.jpg");
                }
            }
        },
        error: function(error) {
            alert("出错了! - "+error);
            $("#loader").delay(500).fadeOut();
        }
    });
    return {rec_movies:rec_movies,rec_movies_org:rec_movies_org,rec_posters:rec_posters,rec_year:rec_year,rec_vote:rec_vote};
}
```

到此为止，整个项目介绍完毕。运行 Flask 主程序文件 main.py，然后在浏览器中输入 http://127.0.0.1:5000/ 显示 Web 主页，如图 10-4 所示。

图 10-4　系统主页

在表单中输入电影名中的单词，系统会实时推荐与之相关的电影名。例如，输入"love"后的效果如图 10-5 所示。

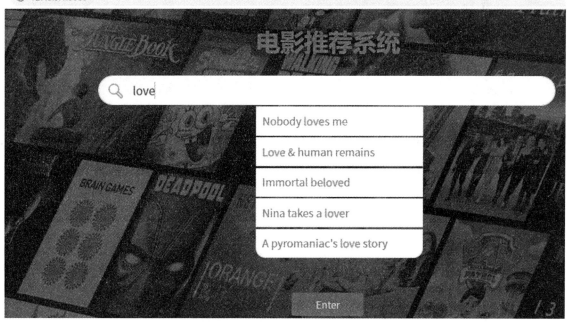

图 10-5　输入"love"后的实时推荐

如果选择实时推荐的第 3 个选项"Immortal beloved"，按 Enter 键后会在新页面中显示这部电影的详细信息，如图 10-6 所示。

图 10-6　电影详情信息